Landforms of the World with Google Earth

Anja M. Scheffers · Simon M. May
Dieter H. Kelletat

Landforms of the World with Google Earth

Understanding our Environment

Anja M. Scheffers
Southern Cross Geoscience
Southern Cross University
Lismore, NSW, Australia

Simon M. May
Institute of Geography
University of Cologne
Cologne, Germany

Dieter H. Kelletat
Institute of Geography
University of Cologne
Cologne, Germany

ISBN 978-94-017-9712-2   ISBN 978-94-017-9713-9 (eBook)
DOI 10.1007/978-94-017-9713-9

Library of Congress Control Number: 2015930258

Springer Dordrecht Heidelberg New York London
© Springer Science+Business Media Dordrecht 2015
This work is subject to copyright. All rights are reserved by the Publisher, whether the whole or part of the material is concerned, specifically the rights of translation, reprinting, reuse of illustrations, recitation, broadcasting, reproduction on microfilms or in any other physical way, and transmission or information storage and retrieval, electronic adaptation, computer software, or by similar or dissimilar methodology now known or hereafter developed.
The use of general descriptive names, registered names, trademarks, service marks, etc. in this publication does not imply, even in the absence of a specific statement, that such names are exempt from the relevant protective laws and regulations and therefore free for general use.
The publisher, the authors and the editors are safe to assume that the advice and information in this book are believed to be true and accurate at the date of publication. Neither the publisher nor the authors or the editors give a warranty, express or implied, with respect to the material contained herein or for any errors or omissions that may have been made.

Cover image: A beautiful syncline in the Andes of NW Argentina, exposing Jurassic rock formations. Large-scale erosion accompanying the uplift during the Andean orogeny has removed the adjacent (anticline) parts of these rock formations, but the trough-like syncline persisted. Picture-perfect cuestas have developed towards the center of the structure, where inclined layers of sedimentary rocks of different colour are incised by draining channels (Image credit: ©Google earth 2012).

Printed on acid-free paper

Springer Science+Business Media B.V. Dordrecht is part of Springer Science+Business Media (www.springer.com)

Preface

Scientific outreach and a passion for the power of observation of the physical world around us are at the heart of this book. From the extraordinary observational skills harnessed by the first scientific travelers in antiquity, to the naturalists and explorers of the eighteenth and nineteenth century like Alexander von Humboldt or Charles Darwin, and through to the modern day, scientists have been the driving force for revolutionary discoveries in the field of earth and environmental sciences. Today, we can observe the surface of the earth when travelling by airplane to every corner of the globe within days or hours, or use virtual globes in the public sphere that are becoming popular and powerful tools to visualize data and information in a geographical context over the ever broadening range of influence of the internet. The patterns, forms and geometries that we can visit with our desktop journey are often of astonishing beauty and staggering aesthetic harmony for the human eye. Besides igniting our scientific quest to understand their origins, they impress us in many psychological and emotional ways as they have impressed artists from all disciplines, painters, sculptors or musicians, over the long time of human history. The exquisite interrelation between arts and sciences and a homage to the power of observation and precise measurements resonates within the words of Alexander van Humboldt: *"Nature herself is sublimely eloquent. The stars as they sparkle in firmament fill us with delight and ecstasy, and yet they all move in orbit marked out with mathematical precision"* (In: "Narrative of Travels of the Equinoctial Regions of the New Continent during Years 1799–1804", London [1814], Vol. 1, pp. 34–35).

Google Earth images showcase the astonishing diversity of the landforms of the world and are the travel tickets to guide the reader along a geomorphologic journey to typical and spectacular landforms in diverse environments on all continents. Google Earth's bird's eye perspective is enriched with photographic images and graphic illustrations and aims to familiarize the reader with diverse terrestrial environments and landforms and the processes that shape them by providing short interpreting texts based on the extensive field experience of the authors. This volume is thought as an inductive addition to existing textbooks on geomorphology, using a language which intends to be understandable for everyone. As the subtitle of our book, "Understanding our Environment" says, we try to provide insights into the diversity of terrestrial landforms especially to young students and scientists and to motivate the interested public to actively observe the landforms and related processes. Part I (Introduction) introduces the reader to the scientific discipline of geomorphology, Part II and III explore the forms of the Earth's surface and the driving forces and processes of nature that result in the landscape and scenery around us. The epilogue touches on the human species as a geologic force in forming and changing the natural environment. Selected short reference lists at the end of each chapter will offer the reader easy access to additional background material that covers the recent progress in the specific topic. An index listing regional and general keywords allows quick searches to special chapters, terms and geographic features.

We thank Google Earth for the permission to use their imagery and are indebted to our editors at Springer Publishing, Petra van Steenbergen and Hermine Vloemans, for their superb support, Frank Schmidt-Kelletat for graphical and technical assistance, and Anne Hager for her support with control and formatting.

Lismore, NSW, Australia	Anja M. Scheffers
Cologne, Germany	Simon M. May
Cologne, Germany	Dieter H. Kelletat

About Google Earth

Virtual, web-based globes such as Google Earth, NASA World Wind or Microsoft Virtual Earth allow all of us to become travelers visiting the most remote places, and tour our planet or even outer space at speeds faster than a rocket. Any computer user can easily, at no charge, download and use Google Earth (for both PC and Mac computers).

If you have not done so already, download Google Earth (the new version) from earth.google.com. Install it on your computer and prepare yourself to fly around the globe on your own research expedition. You can travel to millions of locations and look for the context of all landscape features of interest to you (e.g., geography, geology, vegetation, man-made structures and more). You can also see these objects from different altitudes (i.e., in different scales), perspectives and directions; you can view a chosen area around 360° from an arbitrary point in the air; and you can fly deep into canyons and craters. You can look straight down in a traditional 2D perspective or enable an oblique view in 3D and you can hover above one location, circle around or fly like a bird over countries, continents and oceans. In this book we focus on geologic and geographic features, but that is only a snapshot of what Google Earth is providing with their virtual globe. There is no room here for a complete tutorial, but you will find that the program is so easy to use and understand that you will become an expert after working with it for a few minutes. Please visit the Google Earth webpage for a complete free Google Earth tutorial that is constantly updated to reflect the improvements in different versions of Google Earth (http://earth.google.com/support/bin/answer.py?hl=en&answer=176576).

We hope that the diversity of the landforms of the world will come alive for you and stimulate your curiosity to become an explorer of these fascinating places either as a hobby or profession.

Contents

Part I Introduction

1 Shaping the Surface of Earth: Geomorphology in a Nutshell 3
   Reference 13

Part II Endogenic Forms and Processes

2 Volcanic Landforms 17
   2.1 Volcanic Products 20
      2.1.1 Lava Types 23
      2.1.2 Pyroclastic Deposits 35
   2.2 Types of Volcanoes 36
      2.2.1 Shield Volcanoes 36
      2.2.2 Stratovolcanoes 37
      2.2.3 Lava Domes 37
      2.2.4 Cinder Cones 37
      2.2.5 Calderas 39
      2.2.6 Craters 39
      2.2.7 Volcanic Necks and Diatremes 48
      2.2.8 Hydrothermal Activity 54
   Further Readings 56

3 Igneous Intrusive Landforms 57
   3.1 Plutons 57
   3.2 Dikes and Sills 57
   Further Readings 74

4 Tectonic Landforms 75
   4.1 Folds 80
   4.2 Faults 81
   4.3 Joints 101
   4.4 Circular Structures 101
   Further Readings 120

Part III Exogenic Forms and Processes

5 Impacts Craters 123
   Further Readings 136

6	**Physical and Chemical Weathering**		137
	6.1 Physical Weathering		142
		6.1.1 Insolation Weathering	142
		6.1.2 Frost Wedging and Salt Weathering	142
		6.1.3 Exfoliation	142
	6.2 Chemical Weathering		142
		6.2.1 Hydration, Hydrolysis and Oxidation	142
		6.2.2 Tafoni or Honeycomb Weathering	143
	Further Readings		146
7	**Karst Landscapes: Topographies Sculptured by Dissolution of Rock**		147
	Further Readings		161
8	**Mass Movements: Landforms Shaped Under the Force of Gravity**		163
	8.1 Mass Movement of Hard Rock		164
	8.2 Mass Movement of Unconsolidated Materials		173
	Further Readings		182
9	**Forms by Flowing Water (Fluvial Features)**		183
	Further Readings		244
10	**Planar Forms and Plain Forming Processes: Pediments/Glacis, and Peneplains (with Inselbergs)**		245
	Further Readings		254
11	**Forms by Wind (or: Aeolian Processes): Deflation and Dunes**		255
	Further Readings		291
12	**Glacier Ice and Its Domain**		293
	Further Readings		345
13	**Frost and Permafrost as Morphological Agents (or: the Periglacial Domain)**		347
	Further Readings		374

Part IV Epilogue

14	**Transformation of the Earth's Surface by Man (Anthropogenic Forms)**	377
	Further Reading	387
Index		389

Part I
Introduction

Shaping the Surface of Earth: Geomorphology in a Nutshell

Abstract

Landscapes are shaped by the uplift, deformation and breakdown of bedrock and the erosion, transport and deposition of sediment in a constant cycle of change that operate since the early stages of the formation of our home planet over four billion years ago. The forces of nature that drive these changes originate from processes operating within the Earth's interior or are related to the four inter-connected "geospheres" that comprise the area near the surface of the Earth: the lithosphere, hydrosphere, biosphere, and atmosphere. Every now and then, objects from outer space leave behind their imprints in form of impact craters in the landscape and remind us that Earth is part of the great wonders of the universe. As scientists love to categorize, geomorphologists are not an exemption: We classify landforms as destructive or constructive depending on the forming processes involved. The first chapters introduce the reader to the concepts of geomorphology in a nutshell and explain the forming processes that shape and sculpture our landscape or are unique for our home planet.

The scientific endeavor to understand the nature and history of the forms and the processes sculpturing the surface of the Earth (and other planets) is termed "geomorphology". Terrestrial geomorphology in particular explores the forms and the processes operating upon Earth's landmasses that cover about 30 % of our globe or more than 150 million km^2 as continents and islands. Geomorphologists approach research questions within the rational, empirical and analytical traditions of modern science. This includes observations and information gathering in the field and hypotheses testing in laboratory or field settings or computer modelling. Efforts are taken to replicate and to generalize results. As any other scientists, geomorphologists aim to develop a fundamental framework (or classification scheme) that allows categorizing a certain geomorphic unit of a landscape based on:

- its general structure and shape (*landform*),
- its origin and development (*process*) – geomorphic process types are tectonic, volcanic, gravitational or mass wasting, fluvial, glacial, periglacial, solution, aeolian, or lacustrine and coastal,
- the measurements of its dimensions and characteristics (*morphometry*) such as relief, elevation, aspect, slope gradient, landform width, microfeature relief, dissection frequency or depth, drainage pattern, drainage density and stream frequency,
- and, the presence and status of process overprinting (*geomorphic generation*).

The concept of geomorphic generation incorporates the dimension of time as many forms may have been shaped by different processes over geologic time scales and therefore may have a multi-genetic character. Think of the fjords in Norway or southern New Zealand. The majestic, steep rock faces exhibit rocks that may be hundreds of millions of years old, but the form itself has been initially carved out by surface water in creeks and streams as a fluvial valley several million years ago. Crushing and grinding glaciers transformed these landforms in the typical glacial U-shaped valleys during several glacial periods over the last two million years of the Quaternary and were replaced by fluvial erosion during the warmer climate periods and subsequent marine inundation.

The geomorphic process types can be classified as endogenic or exogenic. Endogenic processes are driven by forces operating in the Earth's interior like mantle convection, plate tectonics, volcanism or earthquakes. These processes leave their footprint in the landscape as endogenic or structural landforms; examples are mountain belts, volcanoes, faults or folds. Exogenic process types operate on or near the surface of the Earth and are associated with one of the four geospheres – the lithosphere, the hydrosphere, the biosphere and the atmosphere. They result in sculptural landforms that are shaped by water, ice and wind (the most important ones) and also life in the widest sense. Biota has a profound influence on weathering and erosion processes and thus is also significant in landscape evolution. Of course, endogenic landforms are under the constant influence of exogenic processes and are shaped and modified by the geomorphic forming agents that govern the past and present environmental conditions.

To us, the joy and inspiration of geomorphology are nurtured by the multitude of possible forming parameters and the detective challenge to unravel the history and evolution of a landscape that we cherish today for its beauty, uniqueness, and intrinsic value or importance as geologic heritage. Each reader can easily comprehend that any landform or topography will depend on a wide variety of environmental parameters – the tectonic and geographical position and associated climatic regimes, the geology in terms of different rock types and the stability of forming conditions over different time scales. Another line of inquiry is whether the landform is an expression of processes operating today as an active dune in the Sahara or whether it is a result of processes operating in the past like the U-shaped valleys as remnants of a glacial world, dry river beds or inactive volcanoes.

The selected Google Earth images in this volume aim to communicate the diversity of geomorphology by presenting a variety of examples for landforms of similar forming conditions. The volume explores this order of process types and starts in Part II with a virtual visit to a wide diversity of endogenic forms originating from processes in the Earth's asthenosphere. Landforms or geological formations that are related to magmatic processes such as volcanism or intrusions of molten rock into the surrounding bedrock are topics in Chaps. 2 and 3. Tectonic processes and their expression as landforms are visualized in Chaps. 4 and II.4. Astonishing as it may sound, the tectonic processes beneath the surface of the Earth have governed the planet's 4.5 billion year evolution and our position in the solar system, where we circle our home star, the Sun, at an average distance of 150 million km, make our Earth unique within the solar system: Earth has a solid surface, an atmosphere in which we can breathe and liquid water we can drink. These three properties allow life, as we know it, to exist.

Plate tectonics are the driving mechanism for the changing geography and steady renewal of the Earth's surface. The outer, solid surface of the Earth is divided into tectonic plates composed of lithosphere, which drifts on top of a 2,900 km-thick mantle, and is constantly moving towards, away from or past each other (Figs. 1.1 and 1.2a, b). The lithosphere is comprised of the Earth's crust, which has a variable thickness of 35–70 km under continents and 5–10 km in ocean basins, and the rigid portion of the upper mantle. The Earth's mantle stores most of the internal heat of the Earth. Flow or convection processes are driven by both thermal and chemical heterogeneities within the mantle. Convection processes in the mantle lead to the formation of volcanic spreading centers under Earth's oceanic crust that stretch across thousands of kilometers through all major ocean basins, i.e. in the form of sea floor spreading along mid-ocean ridges. Here, the continental plates move away from these volcanic spreading centers like conveyor belts and earth scientists term this process plate divergence (Figs. 1.1, 1.2, 1.3, 1.4 and 1.5). The oceanic plates then sink back into the mantle along subduction zones such as off South America or Japan – places where immense geologic forces are evident in the form of earthquakes, mountain building and active volcanoes (e.g., the South American Andes or the Japanese Alps). Around the subduction zones of the Pacific Plate is a region often referred to as the Ring of Fire. It accounts for about 90 % of the world's earthquakes (and 81 % of the world's largest; Fig. 1.2a) and is home to over 75 % of the world's active and dormant volcanoes.

Despite the constant geologic renewal throughout the evolution of the Earth, some landforms or features have a very long history and may have maintained their form and/or position for millions of years. The Finke River in the Northern Territory of the Australian continent is a good example. It is one of the oldest rivers in the world and represents a remnant of a drainage pattern which was active before the Australian plate separated from the Antarctic plate. Australia began its tectonic journey across the surface of the Earth heading north as an isolated continent since ~55 million years ago, and continues to move north by about 7 cm each year. Landforms created by impact craters can be even dated back to more than two billion years of age.

Plate motions are not always smooth. Instead they exaggerate stress along the lithospheric plate boundaries and often plates are stuck together at the edges while the rest of the plates continue to move. As a result, the rocks along the plate boundaries are distorted or what earth scientists call "strained". As the motion continues, the strain builds up to the point where the rock cannot withstand any more bending and breaks. The sudden release of energy in the Earth's crust creates seismic waves that radiate outwards from the rupture and we experience a shaking of the ground – an earthquake. Most earthquakes occur on the boundaries between plates, where one plate is forced under another, in areas off island chains such as Japan, Indonesia or the Solomon Islands, or past another as occurs in California and New Zealand.

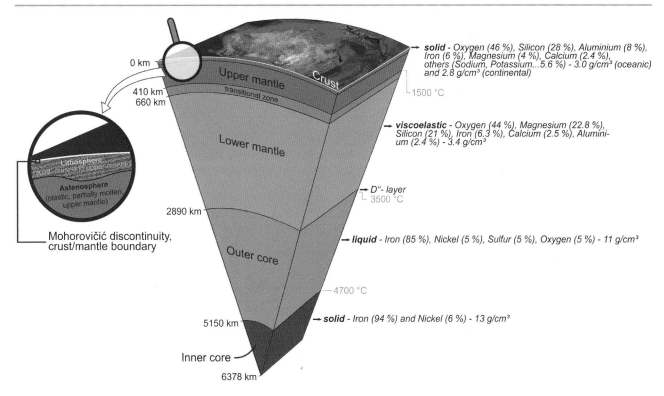

Fig. 1.1 The Earth's interior is composed of four layers, three solid and one liquid (the outer core). The inner core is composed of iron and nickel and temperatures are estimated between 5,000 and 7,000 °C – almost as hot as the surface of the sun. For comparison, the melting point of steel is around 1,500 °C (Image credit: S.M. May, based on Grotzinger et al. 2006)

Earthquakes and long-term tectonic processes can have a visible expression as a feature in the landscape (Fig. 1.7a, b) or at infrastructure level (Fig. 1.8a–c).

Part III opens with a chapter on objects from outer space that leave their imprints in the landscape as extra-terrestrial impact structures. We return to more earthly exogenic matters with the question where landscape materials come from and look at physical and chemical weathering processes. But how do landscape materials get down from mountaintops to valley floors? Gravitational processes or the simple geomorphic transport law "What goes up, must come down" are illustrated in Chap. 8. Planet Earth has a large mass and as a consequence exaggerates significant gravitation on all objects. Any landscape with a topographic gradient will result in the transportation of any material directed downhill, either as mass movements or within a medium such as water or ice. Mass wasting is the geomorphic process by which surface material (soil, unconsolidated sediment, rock) move downslope under the influence of gravity. The movement can occur over different timescales, such as a sudden event that take place within seconds (rockfalls or landslides) or a gradual event that takes years (soil creep).

By far the largest group of landforms is connected to a property of our planet that makes our home in the universe so unique (as far as we know): the existence of liquid water. Fluvial valleys carved by running streams or rivers come in an astonishing variety of sizes, cross profiles or configurations along a river course. Examples for the erosive force of water and the results of accumulation and sedimentation as constructive processes exemplified by alluvial fans and terraces are compiled in Chap. 9. One of the more ordinary "plain" groups of forms in geomorphology (compared to dwindling mountains and gorgeous river gorges) are the plains which are the topic of Chap. 10. As destructive forms (compared to sediment infill of depressions) they are called *peneplains* meaning "nearly flat", as they often are the home of residual single mountains that are consequently called *inselbergs*. The low and featureless slopes along the foothills of mountains and mountain chains are termed *pediments* that translate to "plains at the foot [of mountains]", or *glacis* derived from the cleared area around old fortifications that have been constructed to see any approaching enemy. Wind as a landscape sculptor and its extensive group of landforms and features are illustrated in Chap. 11. The most powerful geomorphic force is associated with glaciers. The weight and movement of these ice masses shape sharp mountain peaks, carve steep U-shaped valleys, whale-shaped bedrock hills and fjords and deposit thick blankets of sediment (gravels or sand) in lowlands and valleys. Glaciers themselves are part of the Earth's surface and they are described visually with all their morphological characteristics in Chap. 12. Landscapes

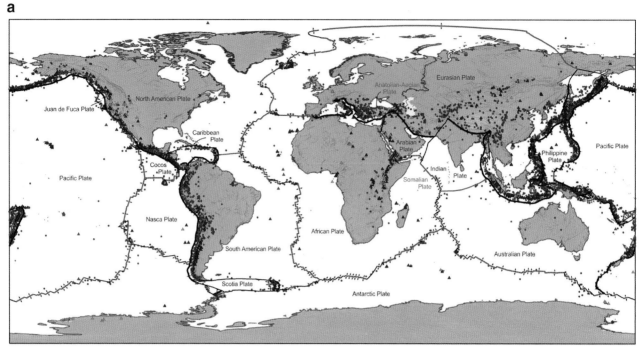

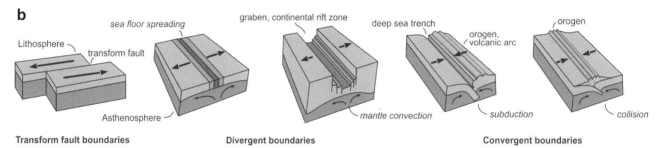

Fig. 1.2 (**a**) The main tectonic plates of the Earth's lithosphere with the three different types of plate boundaries – divergent, convergent and transform boundaries where plates move lateral to each other. The world map shows the distribution of active and potentially active volcanoes. About 80 % are found at boundaries where plates collide, 15 % where plates separate, and the remaining few at intraplate hot spots. Not shown on this map are the numerous volcanoes of the mid–ocean ridge system below the ocean's surface where in fact most of the lava is erupted on Earth's surface (Image credit: S.M. May; based on USGS data and Grotzinger et al. 2006). (**b**) Schematic illustration of different plate boundaries – transform faults or transform plate boundaries, where tectonic plates move laterally along each other; divergent plate boundaries, where new oceanic crust is formed by sea floor spreading or graben and rift structures develop on continents; convergent plate boundaries, where subduction of oceanic lithosphere below oceanic or continental lithosphere or collision of two continental lithospheric plates takes place (Image credit: S.M. May; adapted from Grotzinger et al. 2006)

that depict forms associated with colder climates (permafrost regions) are the focus of Chap. 13.

In Part IV we end our tour through the non-living parts of our environment and briefly look as our own species as a geologic force as can be detected from space in the epilogue. Explanations of the different forming processes are given as introductory remarks (in a nutshell) for the individual chapters.

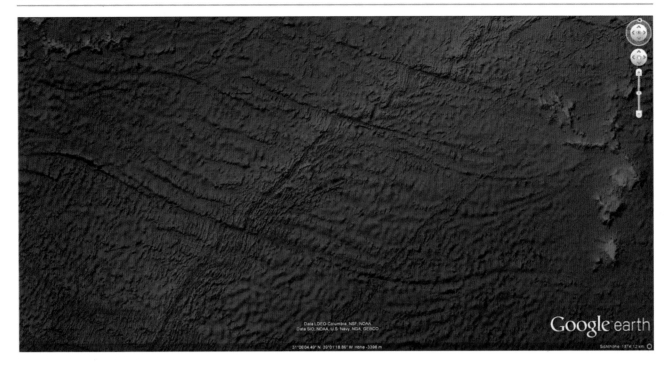

Fig. 1.3 Sea floor spreading along the Mid-Atlantic Ridge generates new oceanic crust: the rigid lithosphere drifts as plates on the asthenosphere. At the line of spreading, a submarine mountain ridge is created, stretching from southwest to northeast in this image. The W-E running transform faults offset the spreading centers of the Mid-Atlantic Ridge, resulting in a step-like appearance. Width of image is ~2200 km. (Image credit: ©Google earth 2012)

Fig. 1.4 (**a, b**) The Mid-Atlantic Ridge is a divergent plate boundary that surfaces above sea level in Iceland, making the rifting process easily visible from space (Image credit: ©Google earth 2012); (**c**) Simple measurement instruments like this metal bar were used to help to estimate the rate of divergence before the era of sophisticated GPS techniques. Iceland has 30 active volcanic systems, of which 13 have erupted since the colonization of Iceland in 874 AD – the last major eruption was by the active volcano Grímsvötn in May 2011 and caused major disruptions in European air traffic (Image credit: D. Kelletat)

Fig. 1.4 (continued)

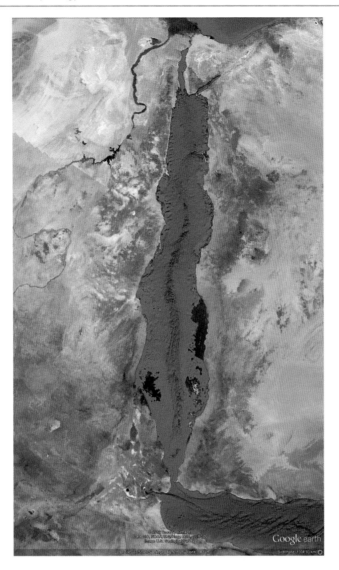

Fig. 1.5 Excellent examples of active spreading centers are the Red Sea and the Gulf of Suez (*upper left*) and of Aqaba (*upper right*). The Red Sea is a young ocean that is forming where Arabia is separating and moving away from Africa. From the southern tip of the Red Sea southward through Eritrea, Ethiopia, Kenya, Tanzania and Mozambique, the African continent is rifting or splitting apart along a zone called the East African Rift. This spectacular geologic event that gives birth to a new ocean will be complete when saltwater from the Red Sea floods the massive rift, probably in several million years from now. The Gulf of Aqaba is continuing in the Dead Sea and the Jordan graben to the north. You may also fly with Google Earth to other spreading zones, such as the Gulf of Baja California in northwest Mexico (Image credit: ©Google earth 2012)

Fig. 1.6 (**a**) The transform boundary along the San Andreas Fault in California (USA) is visible as the straight coastline at Point Reyes National Seashore. It is associated with high seismic activity that produces strong earthquakes such as the San Francisco earthquakes in 1906 and 1989. Riverbeds were shifted for about 150 m (**b**), and even 350 m (**c**) in central California (Image credit: ©Google earth 2012)

c

Fig. 1.6 (continued)

a

Fig. 1.7 (**a**) During the Montenegro or Skopje earthquake in 1963, the bushes shifted for 5 m while a 3 m deep graben formed at the surface. The pasture in the background has been uplifted for nearly 1 m (Image credit: D. Kelletat). (**b**) Over longer geologic time scales, the release of tectonic stress may leave distinct signatures on the landscape such as this 14 km long graben-like structure at the Colorado Plateau (*35°44′N and 111°40′W*; Image credit: ©Google earth 2012)

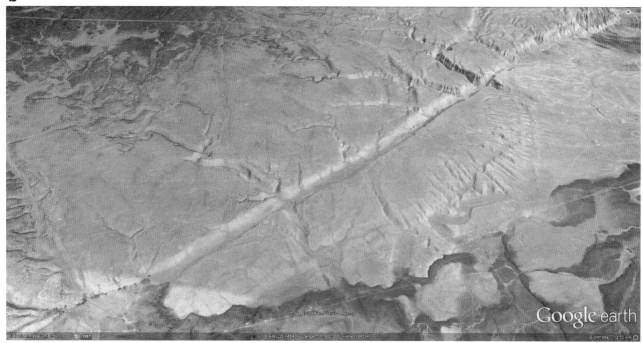

Fig. 1.7 (continued)

Fig. 1.8 (a) A staircase in the Minoan settlement (>3,500 years) of Akrotiri on the island of Santorini/Thira (Aegean Sea, Greece) has been destroyed by compression forces during an earthquake in early historical times. (b) Sideward displacement of the mayor's house in the medieval township of Venzone (northeast Italy) during an earthquake 50 years ago. The small roof tower stayed almost at its old place.

c

Fig. 1.8 (continued) (**c**) In the Nabataean town of Petra (Jordan), the columns of a 2,000-year-old temple have collapsed and are resting in a nice imbricated pattern. The seismic event occurred west of the location (left side of the picture) where the main graben structure of the Dead Sea is situated (Image credit: D. Kelletat)

Reference

Grotzinger J, Jordan TH, Press F, Siever R (2006) Understanding earth, 5th edn. Macmillan, Palgrave

Part II

Endogenic Forms and Processes

Volcanic Landforms

Abstract

Our early ancestors were just spreading out of Africa and east across Asia when one of the most explosive volcanic eruptions on Earth in the last two million years took place at 73,880 years ago (with a margin of error of just a few centuries) in northern Sumatra. The scientific team who dated the event works on the hypothesis that the Toba eruption played a role in shaping human interactions, extinctions and dispersals in Asia and Australia, and has left a legacy of the eruption in our genes. Ancient philosophers were also awed by volcanoes and their fearsome eruptions of molten rock. In their efforts to explain volcanoes, they spun myths about a hot, hellish underworld below Earth's surface and in early Christian society, the idea remained that volcanoes were the gateway to hell. In Roman mythology, Vulcan was the god of fire, volcanic eruptions, and the hearth and forge. He was the gods' blacksmith, making arrows and shields for the deities and whenever a mountain erupted, it was said to be Vulcan pounding on his anvil. The god's legacy remains as the modern name *volcano* is derived from the Latin name "*Vulcanus*", and today means all mountains or hills that are built by lava and other erupted material. But volcanoes do not just occur anywhere, as we shall soon see.

Basically, our ancestors from different cultural traditions around the world had the right idea about Earth getting hotter with depth and that under the surface, the Earth's interior is hot and molten. Today scientists agree that the earth is hot inside due to three main sources of heat in the deep earth: (1) heat from when the planet formed and accreted, which has not yet been lost; (2) frictional heating, caused by denser core material sinking to the center of the planet; and (3) heat from the decay of radioactive elements. Taking the temperature of the inside Earth is made possible by some amazing science and engineering: Over 40 years ago, researchers in the then Soviet Union began an ambitious drilling project, the Kola Superdeep Borehole, whose goal was to penetrate the Earth's upper crust down to 15,000 m. In 1989, a depth of 12,262 m was reached, but the temperatures of 180 °C at this depth and location were much higher than the 100 °C expected by the drilling team. As a consequence drilling deeper was deemed unfeasible and the drilling was stopped in 1992. The temperature profile of the Earth's interior – the geothermal gradient – shows that on average temperatures increase for 3.5 °C per 100 m, but under normal conditions the geothermal gradient is not high enough to melt rocks, and thus with the exception of the outer core, most of the Earth is solid.

Therefore, magmas form only under special geologic settings and consequently volcanoes are only found on the Earth's surface in areas above where these settings occur: To generate magma in the solid part of the Earth either the geothermal gradient must be raised in some way or the melting temperature of the rocks must be lowered. The geothermal gradient can be raised where hot mantle material rises to lower pressure or shallower depth, carrying its heat with it as occurs beneath oceanic ridges, at hot spots, and beneath continental rift valleys. Lowering the melting temperature to partially melt the surrounding rocks and generate magma can be achieved by adding water or carbon dioxide (flux melting) at places deep in the Earth, where the temperature is already high, like subduction zones. Here, water present in the pore spaces of the subducting ocean floor or water present in minerals like hornblende, biotite, or clay minerals would be released by the rising temperature and move into the overlying mantle section to generate magmas.

Magmas are less dense than the rocks that produced them and therefore, as they accumulate, begin to rise upward through the lithosphere. In some places, the melt may find a path to the surface by fracturing the lithosphere along zones of weakness. In other places, the rising magma melts its way toward the surface where we call it lava. The accumulation of lava and other erupted materials as a hill or mountain is called a volcano.

Before the acceptance of Alfred Wegener's plate tectonics theory by the scientific community in 1967, cartographers charted the deep ocean trenches, seismologists plotted earthquakes beneath the trenches, and volcanologists studied the distribution of volcanoes and noted a concentration of volcanoes around the rim of the Pacific Ocean and nicknamed it the Ring of Fire (see Fig. 1.2). Today, the concept of plate tectonics can explain essentially all major features in the global pattern of volcanism such as the observation why so many of the world's volcanoes are situated around the plate boundaries surrounding the Pacific Ocean, where oceanic plates are being subducted beneath the Earth's continental crust. The explanation of the Ring of Fire in terms of plate tectonics and subduction processes was one of the great successes of Wegener's revolutionary theory of continental drift from 1912, although it took nearly half a century to finally convince fellow scientists that Alfred Wegener and later Arthur Holmes were on to something so significant with their theory of plate tectonics that today (another half a century later) is almost regarded as common knowledge by the public.

A world map (Fig. 1.2) shows the distribution of active volcanoes with vents on land or above the ocean surface with the majority occurring in tectonic settings where plates collide (about 80 %) and oceanic lithosphere is subsiding beneath oceanic or continental lithosphere back into the asthenosphere. The remaining volcanoes are found at diverging plate boundaries or occur within plate interiors (Fig. 1.2).

Subduction zone volcanism occurs where two lithospheric plates are converging and one plate containing oceanic lithosphere descends beneath the opposing oceanic or continental plate into the earth's mantle. The lithospheric crust of the subducting plate contains a significant amount of surface water, as well as water contained in hydrated minerals within the basalt of the sea floor. As the subducting slab descends into the Earth's interior, it progressively encounters greater temperatures and greater pressures that cause the descending slab to release water into the overlying mantle wedge. Water has the effect of lowering the melting temperature of the mantle, thus causing it to melt and produce magma that rises upward to produce a linear belt of volcanoes parallel to the oceanic trench. If the oceanic lithosphere subducts beneath another oceanic lithosphere, a chain of volcanoes is forming which contributes to the formation of an oceanic island arc.

Fig. 2.1 The so-called Toba 'super-eruption' created Lake Toba, Earth's largest Quaternary caldera about 100 km long, 30 km wide and 500 m deep is easy to spot with Google Earth in the center of the image. It is estimated that more than seven trillion tons of volcanic materials were ejected, of which at least 800 km^3 was spread as ash across the Indian Ocean and the adjacent landmasses of south and southeast Asia, covering several million square kilometers of the planet's surface in debris (Storeya et al. 2012) (Image credit: ©Google earth 2012)

Modern examples are the Aleutian or Kuril Islands, the Philippine Islands and the Japanese Archipelago (Fig. 2.2a). If the oceanic lithosphere subducts beneath continental lithosphere, then a similar parallel belt of volcanoes will be generated on the continental crust. We call this a volcanic arc and good examples can be found around the Pacific Ring of Fire including the Cascade volcanic arc of the western USA and Canada and the Andes volcanic arc of South America.

Another birthplace of volcanoes is the spreading centers of the Earth's crust, along the mid-ocean ridges (Fig. 2.2b). Most of the lava at Earth's surface that is erupting along these diverging plate boundaries are submarine. Only at certain places we can witness this spectacular geologic event where the spreading centers are surfacing above the ocean such as the volcanic island of Iceland (Fig. 2.2b), or along the graben and rift zone of eastern Africa (Ethiopia, Kenya, Tanzania with the majestic volcanoes of Mt. Kilimanjaro or Mt. Kenya).

Volcanoes may also exist far away from any plate boundaries and geologists struggled very much to explain their existence. Then in 1963, J. Tuzo Wilson, a Canadian geophysicist, provided an ingenious explanation within the framework of the newly accepted plate tectonics theory by proposing the "Hot Spot" or Mantle Plume Hypothesis as we call it today: At some places on Earth, hot magma rises in confined, narrow jets from deep within the Earth's mantle, penetrates the lithosphere and erupts on Earth's surface. These volcanically active hot spots are often visualized as blowtorches anchored in Earth's mantle and are stationary relative to each other. On their epic geologic journey, the lithospheric plates may move over hot spots with the current position of a plate over a hot spot marked by an active volcano. As the plate moves away, the volcano stops erupting and a new one is formed in its place – an island chain is forming. With time, the volcanoes keep drifting away and progressively become older, extinct and eroded. As they age, the crust upon which they sit cools and subsides, and in combination with erosion they eventually submerge below sea level.

The Hawaii Island Archipelago owes its existence to such a hot spot (Fig. 2.3a–c): The main Hawaiian Islands which make up the state of Hawaii (Hawaii, Maui, Oahu, Kahoolawe, Lanai, Molokai, Kauai and Niihau) are the peaks of undersea shield volcanoes that formed at the Hawaiian hotspot, which is presently located under the Big Island of Hawaii. Dating of the volcanoes on each island yields a rate of plate movement of the Pacific Plate westwards over the hot spot of about 10 cm/year with the oldest island above sea level, Niihau (5.5 million years), at the north-western extent of the chain. The outer Hawaiian

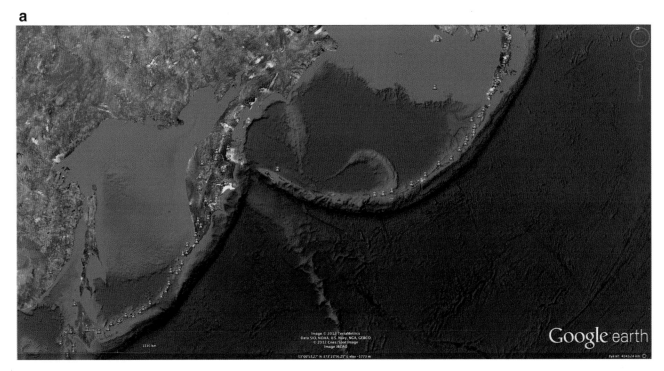

Fig. 2.2 (a) You can see the over 450 volcanoes that make up the Ring of Fire from space, running in a straight line or side by side around the edges of the Pacific Ocean or dominating island chains like Japan, the Philippines or Indonesia. (b) The Mid-Atlantic Ridge, one of the Earth's spreading zones (mid-ocean ridges) stretches from north to south across the entire Atlantic Ocean. It surfaces at Iceland, where high volcanic and tectonic activity (such as graben structures, see Chaps. 1 and 4) manifest the divergence of lithospheric plates (Image credit: ©Google earth 2012)

b

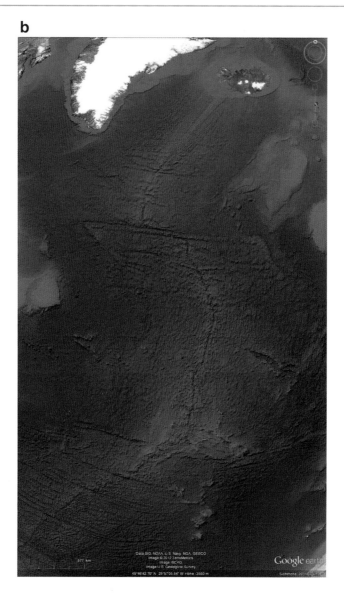

Fig. 2.2 (continued)

Islands are a series of nine smaller, older eroded islands north of Kauai that extend from Nihoa to Kure and represent the above sea level remnants of once much larger volcanic mountains. Even beyond Kure the Hawaiian island chain continues as a series of now-submerged former islands known collectively as the Emperor seamounts (Fig. 2.4).

The endless battle between the forces of nature – creation of new land and destruction by terrestrial erosion and marine abrasion – is very prominent in Polynesian mythology. In Polynesian pre-European mythology, Pele is the fire goddess of Hawaii who was both revered and feared. She is often depicted as tempestuous and destructive, yet beautiful deity, and is said to live in the crater of the volcano of Kilauea on the big island of Hawaii. She could cause earthquakes by stamping her feet and volcanic eruptions and fiery devastations by digging with the Pa'oe, her magic stick. An oft-told legend describes the long and bitter quarrel between Pele and her older sister Namakaokahai, Goddess of the Sea that led to the creation of the Hawaii's volcanic island chain. Tahiti, at the south-eastern end of the Society Islands and the Galápagos Islands are other examples of intraplate volcanism (Fig. 2.5).

2.1 Volcanic Products

The different types and shapes of volcanoes as a landform depend mainly on the chemical composition and the gas content of the magma as it rises through the lithosphere (or lava as it emerges on the land surface) and the rate on which lava is produced, which in turn are principally determined by geologic processes in different plate tectonics settings.

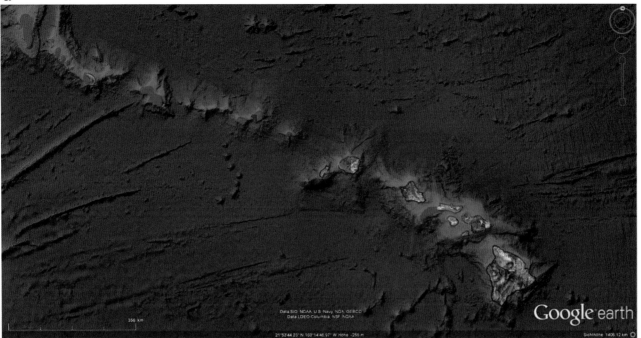
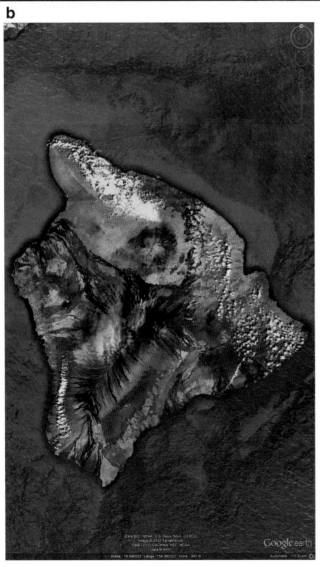

Fig. 2.3 (a) The volcanic island chain of Hawaii in the Pacific Ocean is the result of hot spot volcanism. (b) The largest and highest island, Big Hawaii, is constructed of five major volcanoes: Kilauea, Mauna Loa, Mauna Kea, Hualalai and Kohala. Mauna Loa is the largest active volcano on Earth while Kilauea is presently one of the most productive volcanoes on Earth in terms of how much lava it erupts each year

c

Fig. 2.3 (continued) The youngest volcano, Loihi, is rising steadily from the ocean floor south of Big Island. At present, its summit is 160 m below sea level, but occasionally boiling water and steam clouds makes a spectacular statement about its existence (Image credit: ©Google earth 2012). (**c**) Hawaiian volcanoes primarily erupt a type of lava known as basalt. Basaltic lava is extremely fluid and can flow downhill relatively fast and far, which is why the Hawaiian shield volcanoes generally have gentle sloping sides (Image credit: D. Kelletat)

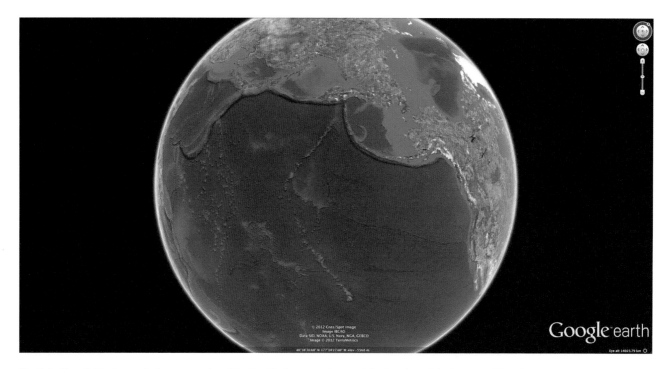

Fig. 2.4 Google Earth reveals the topography of the Pacific Ocean seafloor and the location of the immense Hawaiian Ridge-Emperor Seamount Chain (Image credit: ©Google earth 2012)

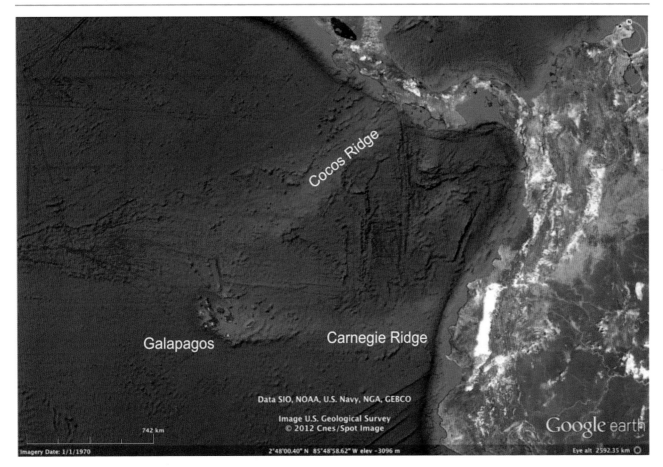

Fig. 2.5 The Galápagos Islands, a group of submarine shield volcanoes, are situated approx. 1,000 km west off the coast of Ecuador within the Nazca Plate. The islands are interpreted to be the result of the Galápagos Hot Spot, currently located beneath the northwestern region of the Galápagos Archipelago near the islands of Fernandina and Isabela. If you hover over the islands with Google Earth, you can see clearly that the Galápagos Islands lie on a submarine platform with an elevation of around 1,000 m below sea level while the surrounding ocean seabed is over 3,000 m deep. The volcanic history of the Galápagos hot spot can be traced by the two adjacent submarine ridges, the Cocos and Carnegie Ridges. The Cocos Ridge extends from the Galápagos Islands to the northeast and subducts beneath Costa Rica. The Carnegie Ridge extends from the islands to the east and is subducted beneath Ecuador. Unlike the Hawaiian Islands, the Galápagos volcanoes do not form a linear chain, but move with the Nazca Plate in an east-southeast direction such that the older islands are found in the southeast. The oldest island is Isla Espanola in the southeast dated to 3.2 million years by isotope dating methods, while Fernandina and Isabela in the northwest are approximately 0.7 million years old, the youngest and most volcanically active. These dates correspond to when the island first surfaced above sea level while the submerged island bases could be up to 15 million years old. The Google Earth image also reveals the subduction process where the oceanic Nazca Plate sinks under the South American Plate: As the descending Nazca Plate plunges downward on the seafloor, it creates a large linear depression called an oceanic trench. These long, narrow deep-sea trenches (visible in the right of the Google Earth image) are the deepest topographic features on the earth's surface (Image credit: ©Google earth 2012)

Very generally speaking, lava usually solidifies into three major types of volcanic (igneous) rocks: basalt, andesite or rhyolite.

2.1.1 Lava Types

Above, we have discussed the broad, shield-like appearance of the Hawaiian volcanoes produced by hot spot volcanism: Intraplate hot spots and divergent plate boundaries such as mid-ocean ridges and continental rift valleys usually produce basaltic magma – the most common magma type on Earth. Basaltic eruptions are rarely explosive and rather occur simply as an overflow when the volcano vent system becomes filled up and lava flows down the flanks engulfing everything in its path. Basaltic lavas have low silica content but erupt at high temperatures (1,000–1,200 °C), and thus are very fluid with velocities commonly reaching a few kilometers per hour and can move over long distances. A basalt flow from Mauna Loa reached a distance of 47 km, which is one of the longest flows ever recorded (see Fig. 2.3b).

In general, the higher the silica content and the lower the temperature, the more viscous the lava is and the more slowly it moves. Basaltic lava flowing on land can depict different surface structures depending on how they cool: Pahoehoe lava (from the Hawaiian word for "ropy") forms when a thin, glassy skin congeals into a thin crust at the surface as the lava cools, but the molten liquid continues to flow underneath that outer skin dragging and twisting it into folds and twists that resemble a rope. The surface is rather smooth (Fig. 2.6a, b) compared to the more viscous a'a flows. A'a lava flows more slowly as the gas content of the lava decreases, forming a thicker skin that breaks up into rougher, sharper blocks and creates a sharp irregular surface that is truly treacherous to walk across (Fig. 2.6c, d). Pahoehoe and a'a lavas can be found in single lava flows and often pahoehoe lava is converted to a'a lava as the lava advances downslope and cooled off during its journey. Figures 2.7 and 2.8 show examples for lava flows of different composition. These lavas solidify to form basalt, which is considered a mafic silicate rock with high contents of magnesium, iron and calcium. Because basaltic lava erupts on the Earth's surface (and often into water) it cools very quickly, and the minerals have very little opportunity to grow. Thus, basalt is commonly composed of very fine crystals of minerals and it is nearly impossible to see individual minerals without magnification.

Pahoehoe or a'a lavas can also form lava tubes, where the lava exposed to the colder air solidifies around a fast-moving liquid lava core. When the lava in that core flows out of the tube, a long tunnel of volcanic rock remains (Fig. 2.9a, b).

The cooling feature of basaltic lavas can create spectacular landforms: As basaltic lava slowly cools and solidifies, it shrinks slightly. The stress causes jointing in the rock in several different planes, and columns of basalt form with a generally hexagonal (6-sided) shape that resembles a handful of pencils. Among the more famous examples of basaltic columns or pavements in the world are the Giant's Causeway in Northern Ireland, Fingal's Cave in Scotland and Devil's Postpile in California (Fig. 2.10a, b).

Fig. 2.6 (a) String-like pahoehoe lava at the base of Mt. Fogo on the Cape Verde islands (Image credit: S.M. May). (b) Perfect examples of pahoehoe lavas are to be found around the active Kilauea volcano on Hawaii. (c) A'a and pahoehoe lavas in close vicinity in the caldera of Fogo Island (Cape Verde Archipelago) (Image credit: D. Kelletat). (d) Typical a'a lava flows with steep slopes from the base of Pico de Teide, Tenerife Island (Image credit: S.M. May)

2.1 Volcanic Products

b

Fig. 2.6 (continued)

c

Fig. 2.6 (continued)

2.1 Volcanic Products

d

Fig. 2.6 (continued)

Fig. 2.7 (a) Several rough old a'a lava flows in the giant caldera of Pico de Teide on Tenerife Island, Canary Islands (Spain). (b) A thick a'a lava flow (>50 m) near Flagstaff in northeast Arizona at *35°40′N* and *111°22′W* is more than 5.5 km long and partly covers a sandstone cuesta. (c) The summit of the 3.718 m high Pico de Teide, on the Canary island of Tenerife (Spain) at *28°16′N* and *16°38′W* is surrounded by a'a lava flows from the last several centuries (Image credit: ©Google earth 2012)

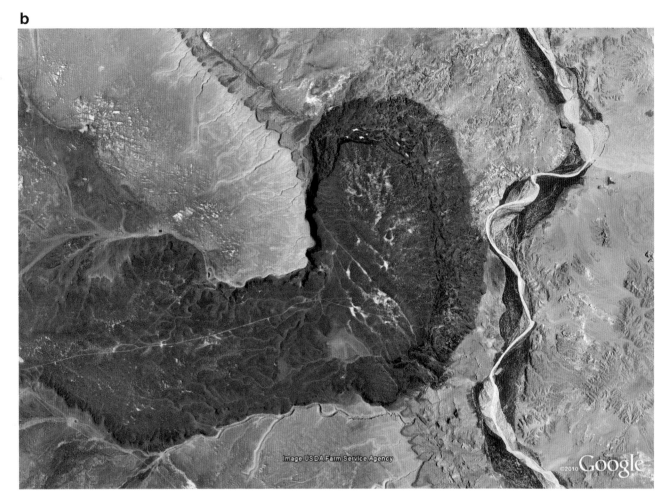

Fig. 2.7 (continued)

c

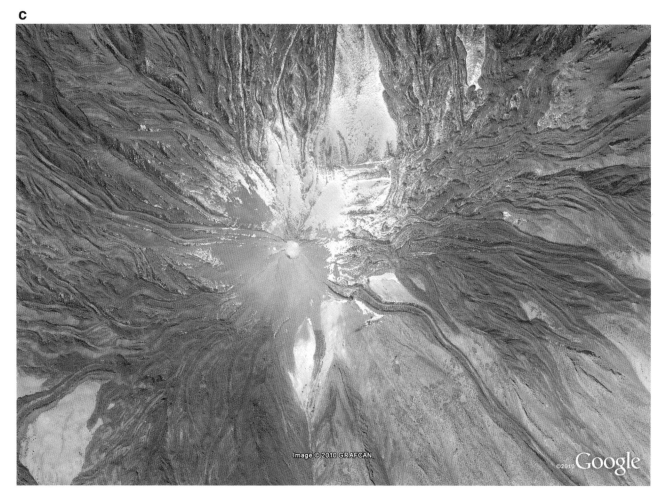

Fig. 2.7 (continued)

2.1 Volcanic Products

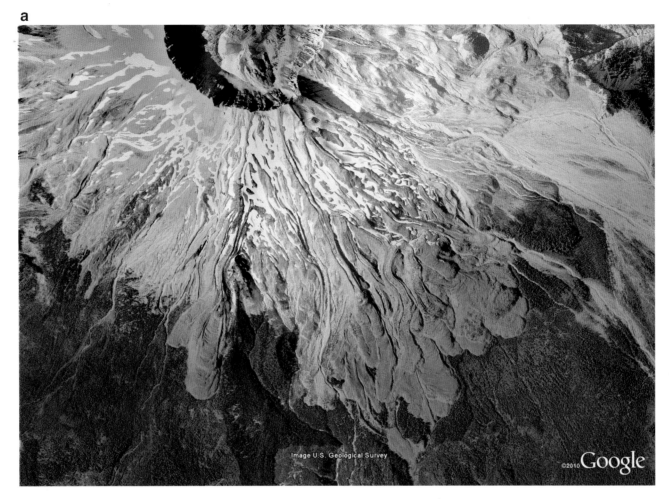

Fig. 2.8 (a) Lava flows on the southern slopes of Mt. St. Helens in southern Washington State, USA. There are a large number of long lava tunnels. (b) Very wide lava flows in central Argentina at *36°23′S* and *69°27′W*. Scene is 51 km wide. The distance covered by a lava flow, the topography (slope) of the mountains, and the thickness and character of the lava (a'a or pahoehoe) allow geologists to estimate its temperature during the eruption, and chemical analyses can be used to infer the depth of formation. (c) Detail of (b), illustrating flow structures within the younger lava. Similar structures are still visible in the older and weathered lava flows lying below. (d) Lava flow from northern Arizona near Flagstaff (Image credit: ©Google earth 2012)

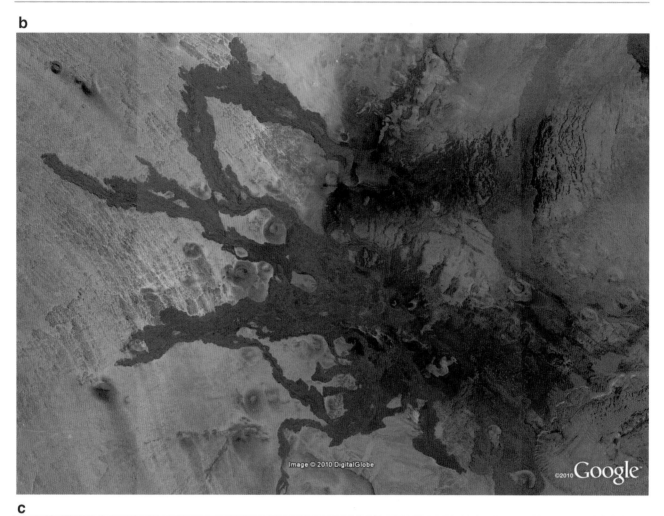

Fig. 2.8 (continued)

d

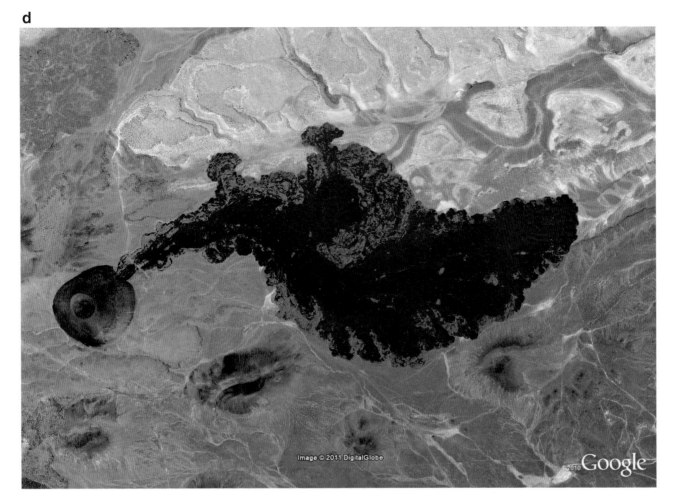

Fig. 2.8 (continued)

Fig. 2.9 (a) A lava tube formed in rough a'a lava at Tenerife Island, Canary Islands, Spain. The outer, rough skin has cooled off rather quickly compared to the inner, finer-grained surface that remained heated while the lava was still flowing through the tunnel (Image credit: D. Kelletat). (b) The Kalkani crater/ash cone in Queensland, north-eastern Australia, at about *18°14′S* and *144°40′E*. In the collapsed lava tube to the south water availability is higher; the lava tube is visible by the green line of dense vegetation (Image credit: ©Google earth 2012)

Fig. 2.10 (a) Vertical basalt columns in southern Iceland, polished by glaciers. (b) The Devil's Postpile, a national monument in the Sierra Nevada of California (USA) showcasing basalt columns of diameters up to 1 m and the height of individual columns up to 20 m. They are in a natural state of destruction due to frost weathering during the cold months (Image credit: D. Kelletat)

Magmas that are produced in the continental mountain belts above subduction zones are more varied in their chemical composition than the basaltic magmas at spreading centers. The most common magma type observed in volcanic arcs is andesitic in composition. Andesitic lavas (derived from the Andes Mountains in South America) move more viscously and slowly compared to basaltic flows and commonly erupt from stratovolcanoes and volcanoes with lava domes. Because of their relatively more viscous nature, they can plug the central vent of a volcano resulting in the build-up of gas beneath the plug. Eventually a violent eruption may blow off the top of the volcano as happened during the 1980 AD eruption of Mount St. Helen, an andesitic volcano in southwestern Washington State. Rhyolitic lavas flow at even lower velocities (often 10 times less than basaltic lavas) and are the most viscous lava with silica content around 70 % and low melting points (600–800 °C). They are produced where large volumes of continental crust have been melted – a good example today is the drifting of North America in the southwest direction over the Yellowstone hot spot, which gave birth to the spectacular and world famous volcanic landscape of Yellowstone National Park. Rhyolite lava exhibits a typical banded structure produced by its flow pattern and in some cases, will cool as glassy obsidian flows.

2.1.2 Pyroclastic Deposits

Amongst the Earth's most spectacular and dramatic geological processes are explosive volcanic eruptions (Fig. 2.11). They eject fragmental material of various sizes from new magma, pre-existing rock and debris from around the vent (Fig. 2.12a). These explosive eruptions occur because the confining pressure of the overlying rocks prevents volatiles such as water or gases in the magma from escaping. Once the magma migrates closer to the surface and the pressure drops, these volatiles release explosive forces shattering all solid surrounding rocks and projecting them into the air. Once the material has been accumulated by gravitational fall, we call them pyroclastic-fall deposits (*pyro* in Greek means

b

Fig. 2.10 (continued)

fire and *klastos* means broken). The generic term tephra is often used to refer to fragmental material of all size classes. Smaller scale eruptions may only reach a hundred meters or so above the vent while powerful explosions inject ash and gases tens of kilometers into Earth's stratosphere, where it can circle the planet for several years and have major impact on global climate. Pyroclasts are classified according to their size, ranging from volcanic ash (<2 mm) to volcanic bombs that can reach the size of a standard family house. Ejected chunks of lava cool during the flight and get their distinct rounded shape of volcanic bombs (Fig. 2.12b). Some of the hot ejected material may weld together and become lithified as tuff deposits (with smaller sized fragments) or as volcanic breccias from larger fragments. Tephra deposits provide valuable time horizons that can be dated to unravel the eruptive history of a volcano and a volcanic region (Fig. 2.13).

2.2 Types of Volcanoes

2.2.1 Shield Volcanoes

Basaltic lava is very fluid and thus spreads easily and widely. Over long periods of time, the gradual build-up of thousands of these flows slowly constructs the characteristically concave, broad profile of a mature shield volcano with relatively gentle slopes that resembles the broad shields used by Hawaiian warriors. Shield volcanoes can vary widely in size with the oceanic shield volcanoes belonging to the largest volcanoes on Earth, but much smaller Icelandic-type shield volcanoes are common in many volcanic regions. In general their volumes can exceed that of stratovolcanoes by several orders of magnitude. The classic example of a shield volcano is Mauna Loa on Hawaii – the largest volcano on Earth (see

2.2 Types of Volcanoes

Fig. 2.11 Explosive ash and cinder eruption at the slope of Mt. Batur on Bali, Indonesia (Image credit: D. Kelletat)

Fig. 2.3). Mauna Loa projects 4,169 m above sea level and is 97 km wide at its base. Hawaii's Mauna Kea is also a shield volcano formed in the same manner, but its profile has been modified by late-stage explosive eruptions, which constructed a series of cinder cones that cap its summit (Fig. 2.3c). Another shield volcano example is the Piton de la Fournaise on the island La Réunion in the Indian Ocean (Fig. 2.14). Shield volcanoes forms where silica-poor basaltic lava is produced in hot spot volcanism or within rift zones (where the crust is diverging). The shield volcanoes in Iceland are constructed along rift zones where the Mid-Atlantic Ridge rises above sea level.

2.2.2 Stratovolcanoes

These enigmatic volcanoes are also known as composite volcanoes because they are composed of stratified layers of both viscous lava flows and pyroclastic material that build the typical concave-shaped profile with steep slopes (Fig. 2.15). Morphology can vary greatly, but this type of volcano typically contains a small summit crater. Stratovolcanoes are common in subduction zones. With their picturesque, distinct morphology and visually dramatic dominance in the landscape but yet violent eruption style they have inspired poets and artists since eons. For example, the Japanese artist Katsushika Hokusai (1760–1849) depicts the classic symmetrical stratovolcano Mt. Fuji in a series of large, color woodblock prints titled the "*Thirty-six Views of Mount Fuji*".

2.2.3 Lava Domes

Lava domes are formed when viscous, silica-rich rhyolitic magma slowly extrudes from a vent and piles up around the vent either as a solitary, domical mass or a complex of overlapping domes. Domes are bulbous, steep-sided structures typically a few tens of meters to a few hundred meters high that can form during single eruptive episodes or by periodic lava extrusion. Lava domes form within the summit crater of volcanoes such as during the 1980–1986 eruption of Mt. St. Helens or on their flanks (Fig. 2.16).

2.2.4 Cinder Cones

Cinder cones, the most abundant of all volcanoes, are built when volcanic vents discharge pyroclasts that accumulate

Fig. 2.12 (a) Several tephra layers on the North Island of New Zealand in the Taupo Volcanic Zone, where most of New Zealand's volcanism of the past 1.6 million years occurred. Taupo itself is a "supervolcano", which erupted 1,800 years ago in the most violent eruption known on Earth in the past 5,000 years. The light-colored layer with pumice clasts has been deposited by this enormous explosion. The eruption plume reached 50 km into the air, with nearby areas buried in more than 100 m of hot pyroclastic flows, which stretched up to 90 km from the vent. It is likely that ash from the eruption was the cause of extraordinary red sunsets recorded by the Romans and Chinese at the time. (b) A large lava bomb (1.5 m across) formed by rotation of hot lava travelling through air during an eruption on Lanzarote Island, Canary Islands, Spain (Image credit: D. Kelletat)

Fig. 2.12 (continued)

with slope profiles that are determined by the angle of repose of its constituent material – the maximum angle at which the fragments will remain stable rather than sliding downhill. Their classic concave-shaped morphology often is characterized by steep slopes and they can be found both as individual features on larger stratovolcanoes or shield volcanoes, or as independent volcanic structures with heights that typically range from a few tens of meters to a few hundred meters. They can form rapidly during single eruptions, but can remain active for longer time periods of several years or decades (Fig. 2.17).

2.2.5 Calderas

Calderas are large, steep-walled volcanic depressions formed by collapse of the summit or flanks of a volcanic structure into underlying magma chambers that have been emptied by very large explosive eruptions or effusion of large volumes of lava. Calderas range from a kilometer to as much as about 100 km in diameter. Many calderas become scenic lakes such as the caldera that forms Crater Lake in Oregon, USA, or when located on islands and inundated by the sea (Fig. 2.18a, b). Large flank collapses of volcanic edifices such as Fogo Island (Fig. 2.19) may result in calderas that are open to one side. In the case of the Fogo collapse, a huge tsunami is assumed to have formed between 80,000 and 120,000 years ago.

2.2.6 Craters

In contrast, craters are typically formed by the explosive ejection of material within and surrounding the upper part of the volcanic vent rather than by collapse. They also differ from calderas in size and are usually much smaller, typically

Fig. 2.13 Landscape to the north of the Puyehue volcano in southern central Chile before (**a**) and after (**b**) the eruption in June 2011. The second image is not a greyscale image as could be mistakenly inferred – it documents layers of volcanic ash covering the landscape and vegetation within a distance of 20 km around the eruptive center (Image credit: ©Google earth 2012)

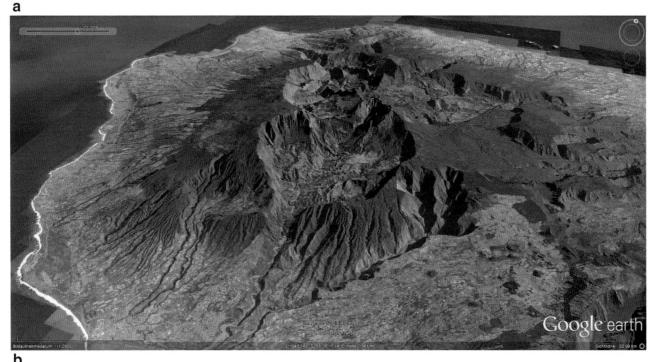

Fig. 2.14 (a) The island of La Réunion in the southern Indian Ocean is divided into two main sections – its northwestern part is characterized by three "cirques", where the inner parts of the older shield volcano have been deeply eroded. (b) Since about 380,000 years, the highly active shield volcano Piton de la Fournaise (2,631 m high) builds up the younger part of the island in the southeast. The volcano complex is characterized by a step-like older and younger caldera. Modern lava flows are mainly concentrated in the caldera flowing into the sea. Very high precipitation supports the incision of deep valleys all around the mountain (Image credit: ©Google earth 2012)

Fig. 2.15 (**a**) The snow- and ice-covered Mt. Osorno (2,652 m above sea level) situated between two lakes in central Chile (*41°07′S* and *72°30′W*, center of image) is an example of a typical stratovolcano of andesitic magmatism. Additional stratovolcanoes visible in the distance are located to the south (Calbuco) and northeast (Puntiagudo) of Mt. Osorno (Image credit: ©Google earth 2012). (**b**) The Mt. Osorno volcano and its typical concave-shaped profile as seen from Lago Llanquihue to the west (Image credit: S.M. May) (**c**) The Mt. Taranaki stratovolcano in New Zealand (2,518 m). (**d**) Mt. Semeru volcano (3,676 m) in eastern Java (*8°07′S* and *112°55′E*, Indonesia) showing recent eruptive activity covering its summit with fresh ashes. The conic shape of strato- and other volcanoes favors the evolution of radial drainage patterns, which is almost perfectly developed here. (**e**) In Argentina, east of the Andean mountain belt, volcanoes may be inactive and partly weathered down by agents of erosion, such as the radial valleys on the giant and lonely Cerro Nevado at *35°34′S* and *68°29′W* with an altitude of 3,820 m (Image credit: ©Google earth 2012)

c

d

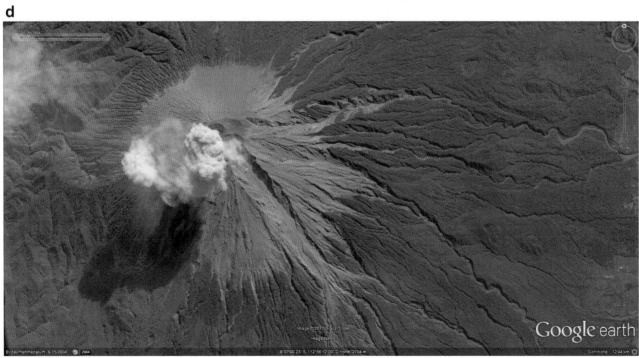

Fig. 2.15 (continued)

e

Fig. 2.15 (continued)

Fig. 2.16 The present crater and central lava dome at the summit of Mt St. Helens (USA). The lava dome formed during a periodic lava extrusion between 1980 and 1986, following the devastating eruption of May 18th, 1980 (Image credit: ©Google earth 2012)

2.2 Types of Volcanoes

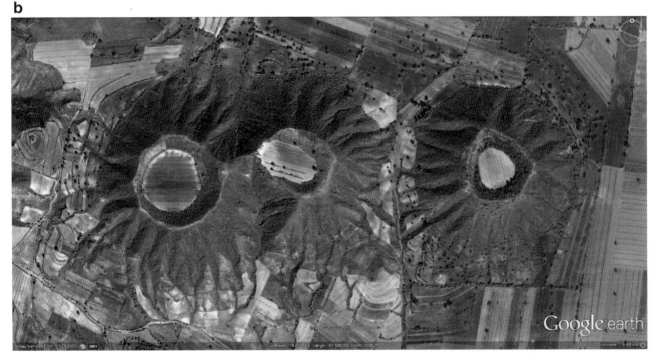

Fig. 2.17 (a) Well-preserved young cinder cones on Lanzarote Island, Canary Islands, Spain (*29°02´N* and *13°43´W*) (Image credit: ©Google earth 2012). (b) Three small cinder cones in central Mexico at around *19°05´N* and *97°31´W*. Width of scene is only 2 km. (c) On the Canary island of Lanzarote (Spain) in the eastern Atlantic Ocean at *29°02´N* and *13°44´W*, long chains of cinder cones along volcanic fissures can be seen. In this image the largest has a diameter of 1.2 km (Image credit: ©Google earth 2012)

c

Fig. 2.17 (continued)

a

Fig. 2.18 (**a**) On the northern Kuriles Islands (Russia) at *49°20′N* and *154°42′E*, a former volcano had collapsed leaving behind a caldera in which a younger volcano starts to form from the old vent within the caldera lake. Width of image is 38 km. (**b**) The 11 km wide caldera of the Santorini island group in the Aegean Sea (Greece) collapsed during a massive eruption in the year 1628 BC during the Minoan period (Image credit: ©Google earth 2012)

2.2 Types of Volcanoes

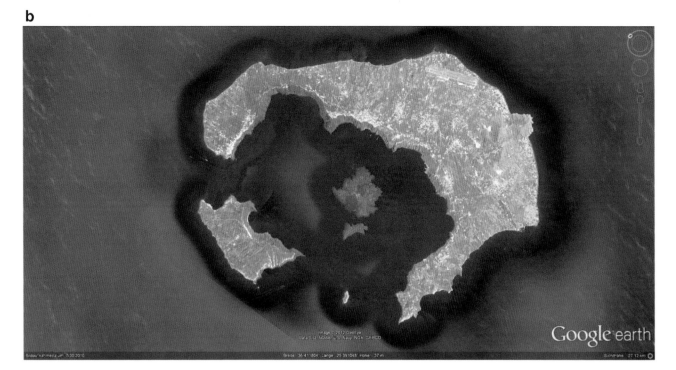

Fig. 2.18 (continued)

Fig. 2.19 (**a**) Ilha de Fogo (Portuguese for fire) is a small island in the Sotavento Island group of the Cape Verde Archipelago (*15°02′N, 24°21′W*) and composed almost entirely of an active volcano. The old volcanic edifice has collapsed and left a caldera of 9 km in diameter. It is surrounded by walls up to 1,000 m high (the Bordeira) and hosts the small community of Chã das Caldeiras (Image credit: ©Google earth 2012). (**b**) The central young summit of Pico do Fogo (2,829 m) in the caldera as seen from the Bordeira, the rim of the caldera (Image credit: S.M. May). (**c**) Erosion along its eastern side gives insight into the volcano's internal stratigraphy with alternating layers of volcanic ash, lapilli and lava. This architecture is responsible for the steep slopes of many stratovolcanoes (Image credit: ©Google earth 2012)

Fig. 2.19 (continued)

defined as being less than 1 km in diameter (Fig. 2.20). Maars and tuff rings are broad craters with surrounding rims that formed during powerful explosive eruptions involving magma-water interaction (Fig. 2.21a–c).

2.2.7 Volcanic Necks and Diatremes

A volcanic neck is an erosional landform often with an irregular, columnar structure that protrudes in bold relief within a landscape of dormant or extinct volcanism. It is created by magma solidifying in the vent of a volcano. Typically, volcanic necks tend to be more resistant to erosion than their enclosing rock formations (Fig. 2.22a).

Diatremes are relatively rare and most unusual, enigmatic volcanic structures (Fig. 2.22b). A famous example is the "Shiprock" in New Mexico, USA which towers at 515 m like a black skyscraper above the surrounding flat-lying sediments of the red, brown desert. Diatremes are only found where gas-charged magma has risen into horizontal sedimentary layers that contain groundwater and reacts explosively. Often the vent and other magma-feeding channels are left behind filled up with volcanic breccia while the surrounding sedimentary layers have been eroded, leaving the diatreme in the landscape. Most sought after are the exotic kimberlite diatremes from very deep mantle-derived magma sources as they often contain precious minerals such as diamonds as is the case in the Kimberley diamond mines in South Africa.

Fig. 2.20 (**a**) A field of explosive craters in the "Phlegraean Fields" northwest of Naples in southern Italy at *40°50′N* and *14°07′E*. The youngest craters date from the seventeenth century and post-volcanic activity such as the exhalation of hot steam and gas as well as mud springs is still present. The scene shown is 10 km wide. (**b**) Explosive maar craters in older basalt layers of northern Patagonia, Argentina. (**c**) The central crater of the Piton de la Fournaise with a diameter of about 1.1 km. Also visible is a smaller caldera and recent activity (Image credit: ©Google earth 2012)

50　　2 Volcanic Landforms

b

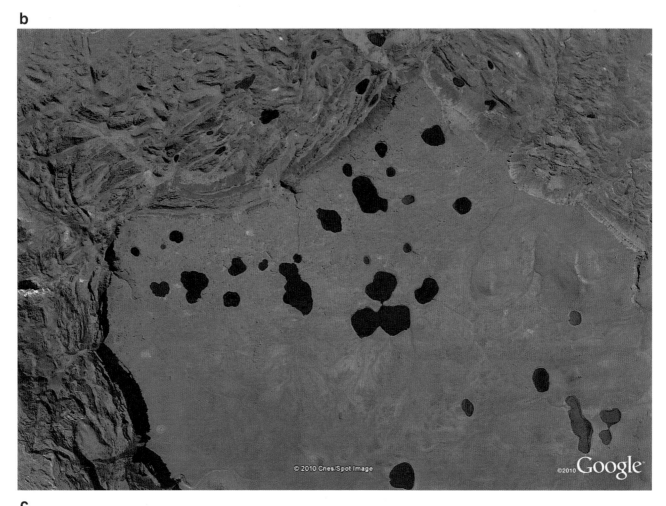

c

Fig. 2.20 (continued)

2.2 Types of Volcanoes

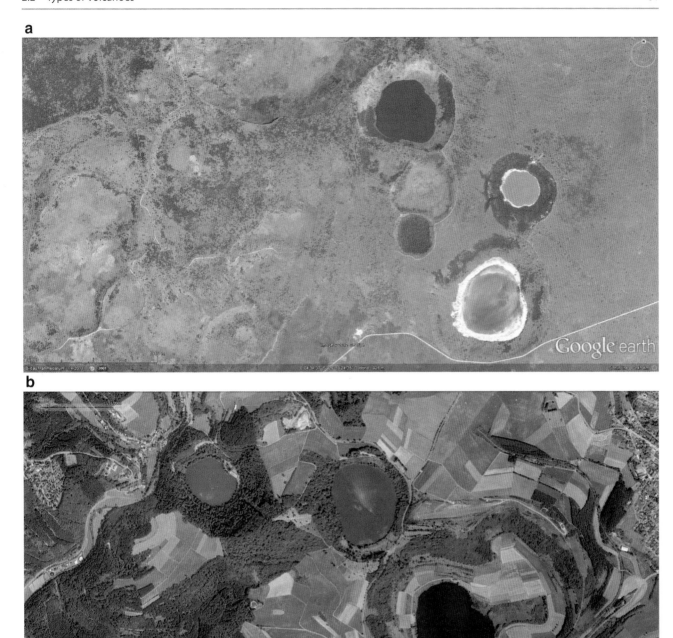

Fig. 2.21 (a) Maar craters in southwest Uganda, Africa. (b, c) The famous Eifel-Maar craters (western Germany) formed by phreatomagmatic volcanism during the mid- to late Pleistocene (b: Gemündener, Weinfelder and Schalkenmehrener Maar; c: Pulvermaar) (Image credit: ©Google earth 2012)

Fig. 2.21 (continued)

Fig. 2.22 (**a**) Devils Tower is a volcanic neck and an enigmatic landmark composed of the uncommon igneous rock phonolite. It was and still is a sacred area for several indigenous Plains Indian tribes and was declared America's first national monument by President Theodore Roosevelt in 1906. It is also considered one of the finest traditional crack climbing areas in North America due to its hundreds of parallel fractures and joints in the rock formation. (**b**) The famous "Ship Rock" (known as *Tse Bitai*, or "the winged rock") in the Navajo country of northwest New Mexico, USA is the remnant central feeder pipe of an explosive volcanic eruption that occurred around 30 million years ago. Ship Rock was probably 750–1,000 m below the land surface at the time it was formed, and has since gained its prominent form due to erosion of surrounding sedimentary sandstone and shale rocks. Another striking feature of Ship Rock is the six dykes that radiate into the landscape from the central, vertical vent. These dikes would have been intruded at some depth subsurface at the time of the eruption (Image credit: ©Google earth 2012)

Fig. 2.22 (continued)

Fig. 2.23 (**a**) A fumarole at Nea Kaimeni in Santorini, Greece producing yellow sulfur crystals. Fumaroles are basically steam vents that allow water vapor and gases to escape on the surface of the Earth – on land or on the floor of the ocean. (**b**) Boiling mud spring in the Waiotapu valley of New Zealand's North Island. Generally, mud pools are very acidic hot springs that dissolve the surrounding rock, turning it into finer particles of clay and silica that become suspended in the water. When the hot water and steam rises from below, it forms bubbles that burst when they reach the surface and push water and sediment to the edges of the pool where the sediment builds a mound that make the opening look like a crater (Image credit: D. Kelletat)

b

Fig. 2.23 (continued)

2.2.8 Hydrothermal Activity

Volcanic activity can continue in the form of hydrothermal activity for decades or even centuries and millennia after the eruption of lava or pyroclastic material has ceased. Steam or gas emissions through small vents (fumaroles) are surface manifestations of the circulation of water through subsurface hot volcanic rocks or magma, which produces hot springs and geysers when the heated water returns to the surface (Figs. 2.23 and 2.24). Hot springs vary in temperature and can be calm, effervescent, or boiling depending on temperature and distance of the magma below. When the heated water travels up to the surface, it dissolves material from the surrounding bedrock and carries this material with it during the ascent. This is the reason why hot springs tend to be rich of minerals, and people often used these hot springs for medicinal purposes over centuries. Many of these geothermal features are very colorful due to mineral deposition, but minerals are not responsible for all of the colors. For a long time it was believed that life forms could not exist due to the extreme heat and high acidity of many hot environments, but researchers discovered that microorganisms known as thermophiles (literally "heat loving") can live and actually thrive in these very hot hydrothermal waters.

2.2 Types of Volcanoes

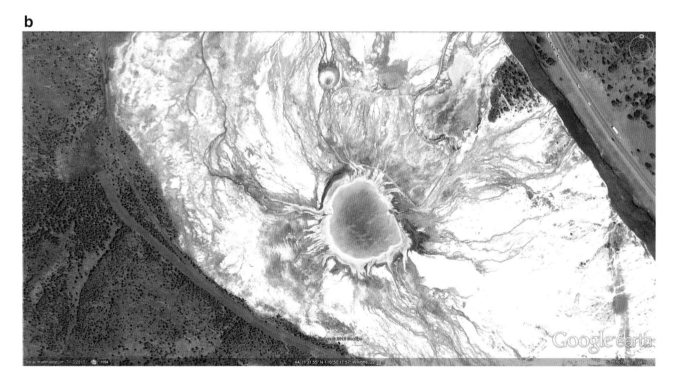

Fig. 2.24 (a) At Mammoth Hot Spring in Yellowstone National Park of Wyoming, USA, dissolved minerals, i.e. "geyserite", crystallize when hot geothermal water evaporates. This geyserite forms large pools with basins, rims and cascades. The pools in the image are less than a 1,000 years old and undergo constant change (Image credit: D. Kelletat). (b) Hot springs in Yellowstone National Park, USA, show an intriguing color pattern: At the center of the pool, the temperature is too high for life to thrive. Below about 94 °C yellow and orange carpets of bacteria start to grow and in cooler areas of about 76 °C dark algae surround the pool. The Grand Prismatic Spring with a diameter of 80 m (*44°31′ 30″N* and *110°50′17″W*) is the largest hot spring in Yellowstone National Park (Image credit: ©Google earth 2012)

Further Readings

Bardintzeff JM, McBirney AR (2000) Volcanology. Jones & Bartlett Publishers, Sudbury

Bryan TS (1995) The geysers of Yellowstone, 3rd edn. University Press of Colorado, Boulder

Chesner CA, Westgate JA, Rose WI, Drake R, Deino A (1991) Eruptive history of earth's largest quaternary caldera (Toba, Indonesia) clarified. Geology 19:200–203

Day SJ, Heleno da Silva SIN, Fonseca JFB (1999) A past giant lateral collapse and present-day flank instability of Fogo, Cape Verde islands. J Volcanol Geotherm Res 99:191–218

Decker R, Decker B (1997) Volcanoes. WH Freeman & Co, New York

Diefenbach AK, Guffanti M, Ewert JW (2009) Chronology and references of volcanic eruptions and selected unrest in the United States, 1980–2008. US Geological Survey, Open-File Report 2009-1118. Reston, Virginia (USA)

Fisher RV, Heiken G, Morris AK (eds) (1998) Volcanoes. Crucibles of change. Princeton University Press, Princeton

Gates AE, Ritchie D (2007) Encyclopedia of earthquakes and volcanoes. Checkmark Books, New York

Krafft M (1993) Volcanoes. Fire from earth. Harry N Abrams, New York

Marti J, Ernst G (2005) Volcanoes and the environment. Cambridge University Press, Cambridge

Oppenheimer C (2003) Climatic, environmental and human consequences of the largest known historic eruption: Tambora volcano (Indonesia) 1815. Prog Phys Geography 27:230–259

Paris R, Giachetti T, Chevalier J, Guillou H, Frank N (2011) Tsunami deposits in Santiago island (Cape Verde archipelago) as possible evidence of a massive flank failure of Fogos volcano. Sediment Geol 239:129–145

Sigurdsson H (1999) Melting the earth: the history of ideas on volcanic eruptions. Oxford University Press, Oxford

Sigurdsson H, Houghton B (eds) (1999) Encyclopedia of volcanoes. Elsevier, Amsterdam

Storeya M, Roberts RG, Saidinc M (2012) Astronomically calibrated 40Ar/39Ar age for the Toba supereruption and global synchronization of late Quaternary records. Proc Natl Acad Sci. doi:10.1073/pnas.1208178109

Igneous Intrusive Landforms

Abstract

If you come across igneous rocks consisting of coarse-grained minerals in a landscape, chances are you are standing on an igneous intrusion that crystallized several kilometers below the Earth's surface. Igneous intrusions are fascinating windows into magmatic processes that take place deep in the Earth's crust and cannot be observed directly with our current scientific methods at hand. But igneous intrusions that have been uplifted and exposed by erosion such as plutons and dikes provide insights about magmatic processes that took place millions of years ago and we can deduce the size, shape and magma composition by studying their geology exposed at the Earth's surface. Plutons are fundamental building blocks of the continental crust and are often composed of the oldest rocks in Earth's history, while dikes represent the fossil remains of the "volcanic plumbing system".

3.1 Plutons

Plutonic rocks are named after Pluto, often considered King of the Underworld in Roman mythology. He was worshipped as the god of riches and wealth from under the Earth and igneous intrusions often contain deposits of precious ores. Plutons are large igneous bodies that intrude Earth's crust at depths of 8–10 km, where they cool slowly over geologic time allowing larger crystals to grow from the melt. The cooling time of a pluton depends on its depth and mass and may exceed millions of years. Plutons may represent the magma chamber of an extinct volcano or a magma body that never produced any eruptions. Usually plutons are composed of coarse-grained granite rocks that stand out in a landscape because of the lighter colors (near white to gray to pink) compared to the darker basalts of extrusive volcanic origin. The shape and size of plutons are highly variable ranging from one to hundreds of cubic kilometers. Most intrusions show a sharp zone of contact with the surrounding rocks that gives evidence that liquid magma intruded solid bedrock (Figs. 3.1, 3.2, 3.3, 3.4, 3.5, 3.6, 3.7, 3.8, 3.9 and 3.10). The largest plutons cover at least 100 km^2 and are called batholiths. It is thought that they indicate a long period of repeated igneous intrusions over a large area, as might be expected along a subduction zone.

3.2 Dikes and Sills

Dikes are similar to plutons in many ways but have a different relationship to the surrounding rocks (Figs. 3.11 and 3.12). They represent pathways and routes of magma transport through the volcanic plumbing system in the Earth's crust where the melts follow any pre-existing and new cracks that open up due to the pressure of the rising magma. Dikes solidify as sheet-like igneous bodies that show a discordant relationship to the rocks in which they intrude. Discordant means that they cut across pre-existing structures. Their size and shape can vary from a few centimeters to many meters and sometimes, they can be traced for several kilometers through a landscape. Most often, they occur in swarms of hundreds and thousands of dikes of different size classes in a region that has been intruded by large igneous bodies. Sills are smaller, tabular intrusions (<50 m thick) that have intruded between layers of rock but do not cut across the units. Usually, sills are fed by dikes, but these may not be exposed on the surface.

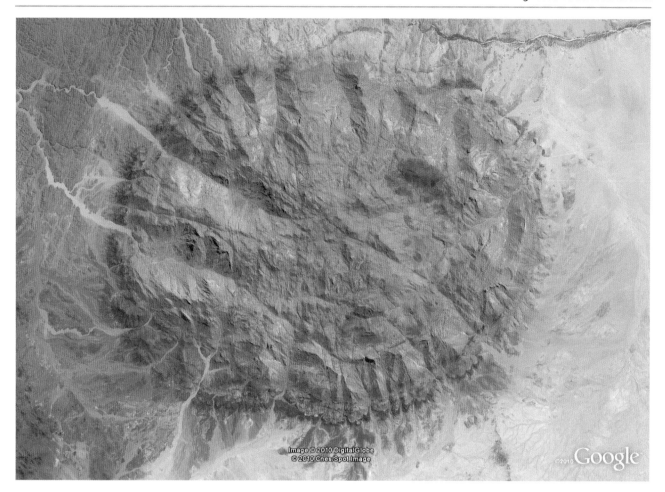

Fig. 3.1 The Brandberg with a diameter of 45 km in northern Namibia (southwest Africa) is a perfect example of a granitic intrusion (*21°06′S, 14°32′E*). It towers 1,600 m above the surrounding plains. Radial joints and in particular the outer ring of darker rocks demonstrate that magma made its way by melting and heating the surrounding country rocks and changed their mineralogy and texture by a process called contact metamorphism. In fact, this darker ring of rocks is differentiated into several mineral assemblages reflecting the temperature gradient from the inner to the outer parts: when intrusions cool they will crystallize the fastest where they are in contact with the colder country rocks. Crystals at the edge of the intrusion will be smaller than those in the center. The outside edge with the smallest crystals is called the chilled margin. The country rock closest to the intruding magma may also recrystallize, forming the baked margin (Image credit: ©Google earth 2012)

3.2 Dikes and Sills

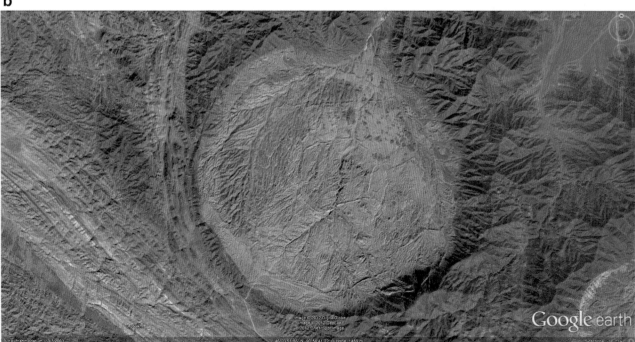

Fig. 3.2 (a) Plutonic intrusions in southern Mongolia (*42°16′N, 105°43′E*). (b) One of many plutons in the Gobi-Tienshan Intrusive Complex near the western border of Mongolia (Image credit: ©Google earth 2012)

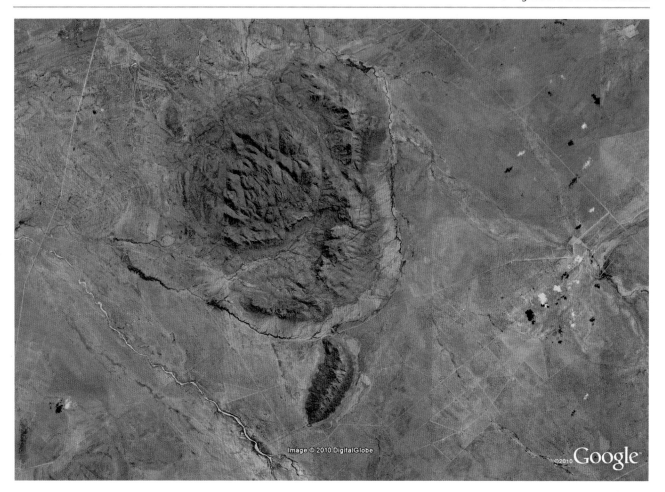

Fig. 3.3 A nearly circular intrusion in northern Namibia (southwest Africa) (*20°22´S, 16°15´E*). Contraction during the cooling process formed the jointing pattern observed (Image credit: ©Google earth 2012)

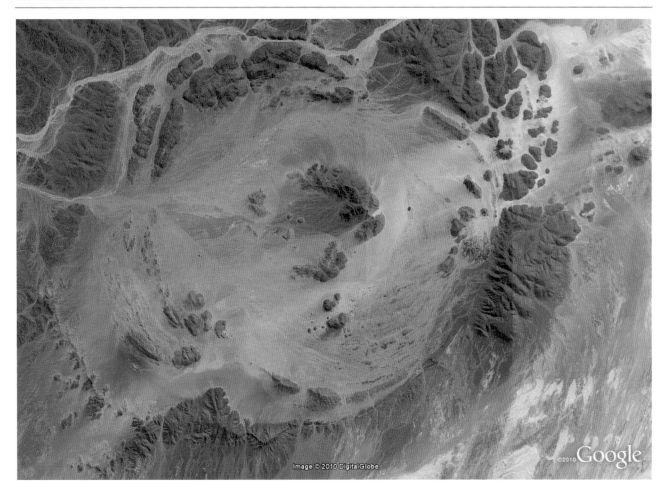

Fig. 3.4 From this intrusion in western Namibia (southwest Africa) only the ring structure of the hard metamorphic rocks is visible or preserved (*21°24′S, 14°12′E*) (Image credit: ©Google earth 2012)

a

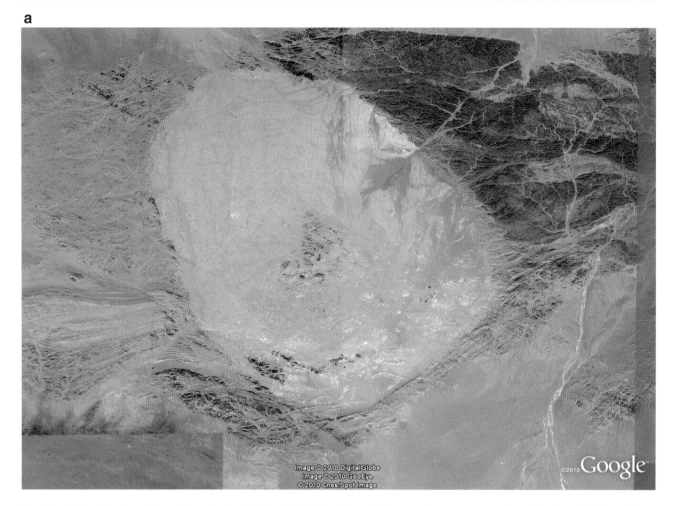

Fig. 3.5 (**a**) An intrusion at the northern border of China (*42°10′N, 101°48′E*). The pluton has a maximum diameter of 15 km. (**b**) A compound pluton in northern China at *42°0′N* and *90°24′E*, where the longest diameter in southwest-northeast-direction is 30 km (Image credit: ©Google earth 2012)

b

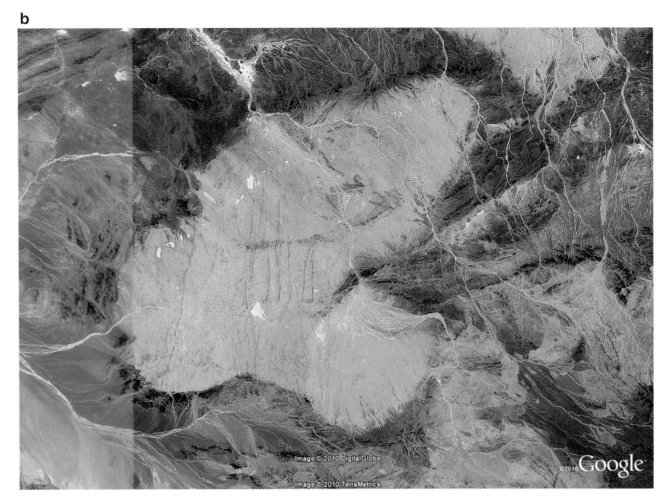

Fig. 3.5 (continued)

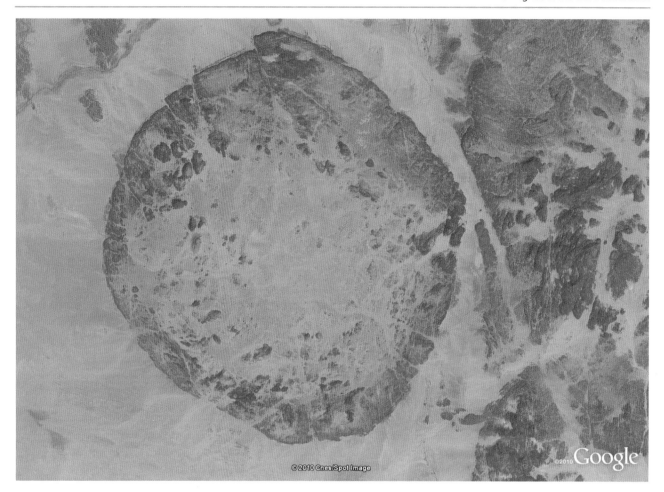

Fig. 3.6 Pluton in southern Algeria at *18°28′N* and *1°07′E* (Image credit: ©Google earth 2012)

3.2 Dikes and Sills

Fig. 3.7 (**a**) With an elevation of 200 m (above surface), ~750 m length and 500 m width, the Bald Rock (Bald Rock NP) is the largest exposed granite-type monolith (Stanthorpe Adamellite) in Australia. It belongs to the New England granite belt of northern New South Wales and southern Queensland (Australia), formed during the Triassic period by intrusive magmatism. Later, the granitic complex was exposed by erosion of less resistant surrounding rocks (Image credit: ©Google earth 2012). (**b**) View from the top of the Bald Rock to the northeast. Note the typical smooth surface and round forms, exfoliation (image center) as well as the characteristic boulders, resulting from corestone-type granite weathering (see Chap. 6) (Image credit: S.M. May)

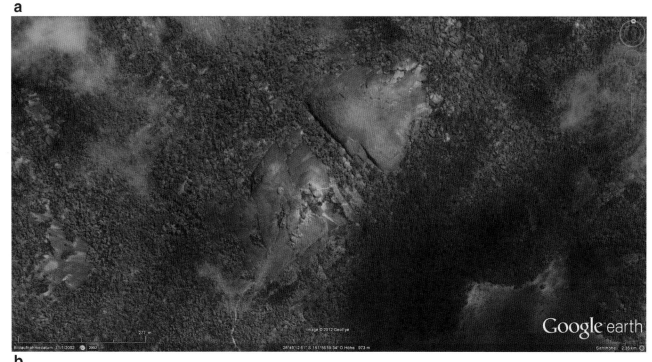

Fig. 3.8 (a) Similar to the Bald Rock monolith, the so called Pyramid (Girraween National Park, Australia) and its neighboring granitic outcrop (b) belong to the New England granite belt as well (Image credit: ©Google earth 2012). (b) View to the north as seen from the Pyramid, exemplifying the dome-like form of numerous granitic outcrops in granite landscapes (Image credit: S.M. May)

3.2 Dikes and Sills

Fig. 3.9 (a) Joshua Tree National Park, California (USA), one of the most famous granite landscapes of the world. Different zones of the intrusive complex (image center, *light grey* and *brown grey* color) and the contact zone of the magmatic intrusion and surrounding rocks (*left part* of image) shown in (c) and (d) (Image credit: ©Google earth 2012). (b) Granite detail in the Joshua Tree National Park (Image credit: S.M. May). (c) Contact zone of the magmatic granite intrusion and surrounding rocks (Image credit: ©Google earth 2012). The contact area is characterized by contact metamorphism (dark brown rocks) (d) Photograph of the same location as in (c) (Image credit: S.M. May)

Fig. 3.9 (continued)

Fig. 3.10 (a) Vertical intrusive dike in the "Bordeira", the up to 1,000 m high caldera walls on Fogo, Cape Verde islands. (b) A different smaller dike close to the dike illustrated in (a), showing columns that display cooling directions, similar to basaltic columns (Chap. 2), (Image credit: S.M. May)

Fig. 3.10 (continued)

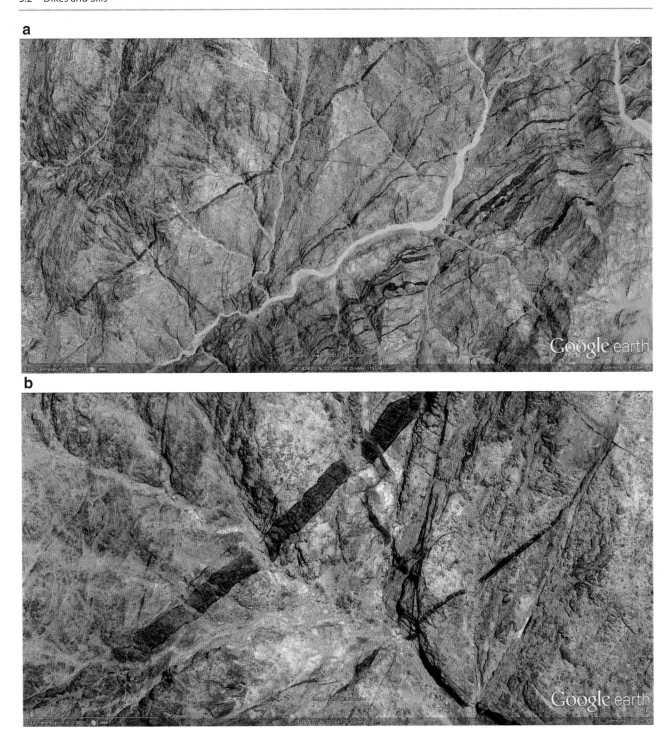

Fig. 3.11 (a) A swarm of basalt dikes in the northern Sinai Peninsula, Egypt, at *28°18′N* and *33°56′E*. (b) Detail of (a), showing at least two different generations of dikes crossing perpendicularly. Dark outer boundaries of the massive dike illustrate the contact zone to the surrounding rock (Image credit: ©Google earth 2012)

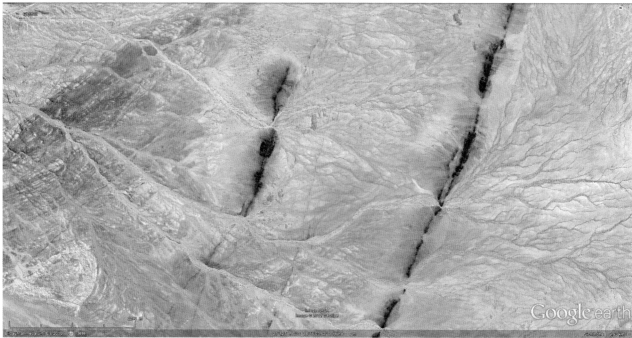

Fig. 3.12 (a) Parallel dikes in southwest Angola (Africa) at *15°49′S, 12°15′E*. (b) Two parallel, distinct and resistant dikes in Namibia. Erosion and weathering is indicated by darker color around the dikes; *21°42′S, 14°58′E*. (c) A swarm of dikes in China, at *43°04′N, 97°26′E*. (d) Dikes in southwestern Sinai. Due to their higher resistance to weathering, they are preserved in a typical straight wall-like form, *27°55′N, 34°02′E* (Image credit: ©Google earth 2012)

3.2 Dikes and Sills

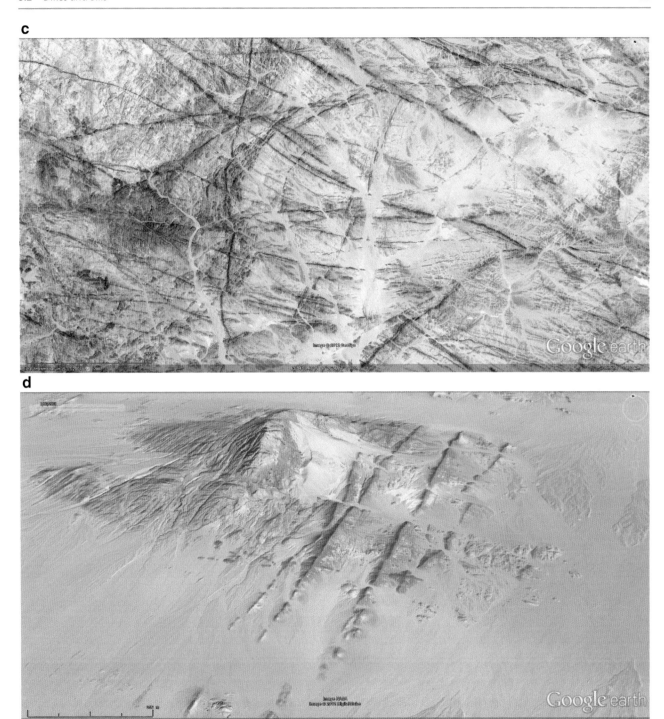

Fig. 3.12 (continued)

Further Readings

Grotzinger J, Jordan TH, Press F, Siever R (2006) Understanding earth, 5th edn. Palgrave Macmillan, London

Miller JS (2008) Assembling a pluton…one increment at a time. Geology 36(6):511–512

Press F, Siever R (1994) Understanding earth. WH. Freeman & Co., New York

Tectonic Landforms

4

Abstract

Why is the Earth so restless and relentlessly changing? And what forces can tilt, bend or fracture rocks that seem so rigid and strong into wild and amazing patterns of folds, faults or fractures? Early geologists who understood all too well that most sedimentary rocks had been laid down as soft, horizontal layers have wrestled with questions such as these for centuries. Most educated Europeans during the medieval age until the 1700s were convinced that a biblical flood played a major role in shaping the Earth's surface and geologic change is caused by a series of catastrophes. This way of thinking was known as "catastrophism" until James Hutton, a Scottish geologist, proposed a new way of thinking in 1785 centered on the "Uniformitarian Principle" which stated simply and elegantly: *"The present is the key to the past"*. This concept would become the spark that ignited a new way of viewing the Earth. It assumes that the geologic forces and processes – gradual as well as catastrophic – acting on the Earth today are the same as those that have acted in the geologic past. Pivotal in this paradigm change was the observation of tectonic forces that can indeed tilt ancient sediments. At Siccar Point (today famous in the history of geology), Hutton observed an angular unconformity where the gently sloping strata of the 345 million year old Devonian Old Red Sandstone overlie near vertical layers of 425 million year old Silurian greywacke. This chapter explores briefly the world of tectonic stress and its associated landforms of faults, folds, joints, domes and basins.

Rocks will respond to tectonic forces by deformation. Deformation is a general term that describes the compression, extension, shearing or folding and faulting of rocks by plate tectonic forces. The discovery of plate tectonics in the 1960s established that the Earth's outermost layer, the lithosphere, is fragmented into approximately a dozen large and small plates that are moving relative to one another as they ride atop hotter, more mobile material of the asthenosphere. Plate tectonic forces vary depending on the direction of plate movements relative to each other: Tensional forces, dominant at divergent plate boundaries, stretch and pull geologic formations apart. The associated geologic landform are rift valleys – long, narrow troughs that form when a block of rock sinks downward relative to its two flanking rock formations along nearly parallel, steeply dipping faults. Classic examples include the rift valleys of East Africa (Fig. 4.1), the Red Sea, the Rio Grande Valley, the Jordan Valley in the Middle East and any rift valleys at mid-ocean ridges.

At the rift valley system in East Africa, Somalia, Dschibouti, Eritrea, Ethiopia, Kenya, Tanzania and a part of Mozambique are moving towards the east, while the rest of central and western Africa remains more or less *in situ*. The bottom of the rift valley is characterized by a chain of very large and deep lakes. Within the next 10 million years, East Africa will probably be separated from the continent and an arm of the Indian Ocean may fill the rift that runs from the Red Sea to south of Madagascar. Figures 4.1 and 4.2 show tectonic graben structures (a tectonic trough) in this large rift system.

Fig. 4.1 (a) A section of the East African Rift valley in Ethiopia. The breaking of crust sections and vertical movements occurs along major faults and shoulders of the rift valley and is clearly visible. Tensional forces along the diverging crustal blocks result in the subsidence of the inner parts of the blocks, forming a basin or depression (tectonic graben) with increased sedimentation (*10°53′N* and *41°22′E*). (**b**) Detail of (**a**), showing the inner part of the rift zone. In the subsided inner parts of the rift valley, sediments accumulate (note the alluvial fans) (Image credit: ©Google earth 2012)

Compressive forces characteristic for convergent boundaries will squeeze and shorten rock formations. Along transform fault boundaries, where plates slide past each other, shearing forces will push two parts of a rock formation in opposite directions. These forces, caused by tectonic stress, not only cause deformation directly along plate boundaries, but also can be transferred for hundreds or thousands of kilometers into the plate's interior because continental crust does not

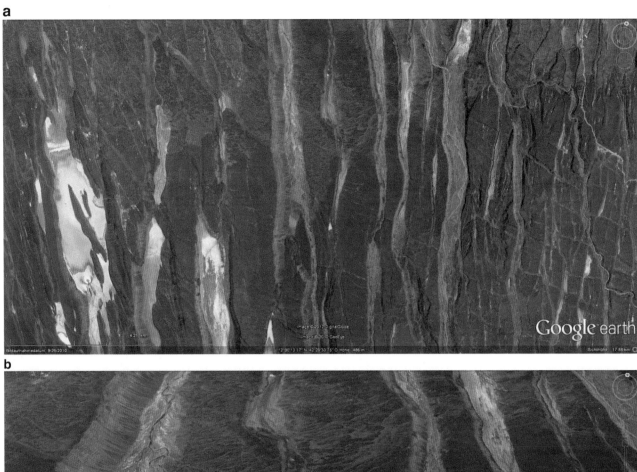

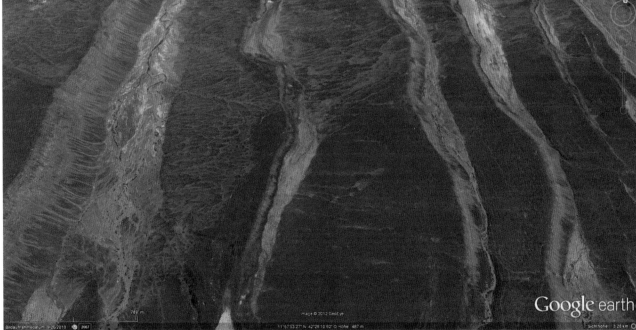

Fig. 4.2 (a) A sequence of graben and horsts in Dschibouti, part of the East African Rift (*11°58′N* and *42°29′E*). (b) Detail of (a) (Image credit: ©Google earth 2012)

behave rigidly within these broad zones. The World Stress Map (Fig. 4.3) illustrates that principle well and aims to characterize the Earth's contemporary tectonic stress patterns, as well as to understand the stress sources. Whether a rock formation responds to these forces by folding or faulting or a combination of both depends on the orientation of the forces, the rock type and other physical conditions (temperature, depth and pressure) during the deformation process. Geologists refer to forces as stresses (the force per unit area) and the response in terms of deformation is a form of strain.

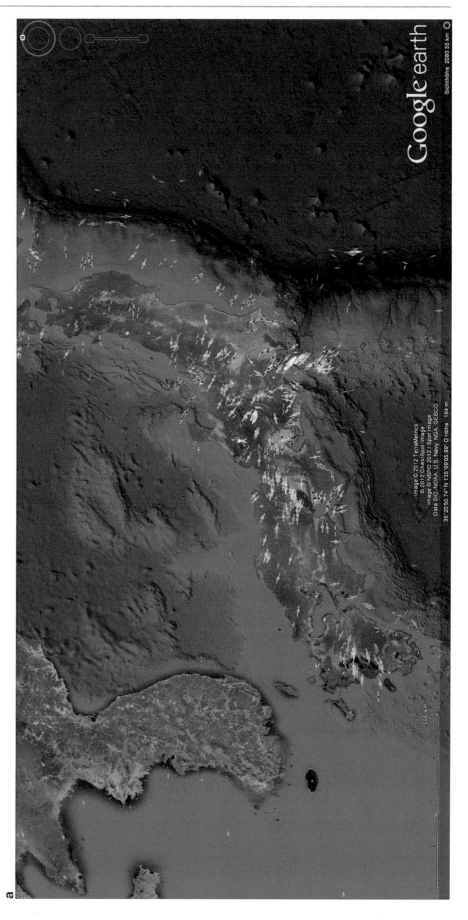

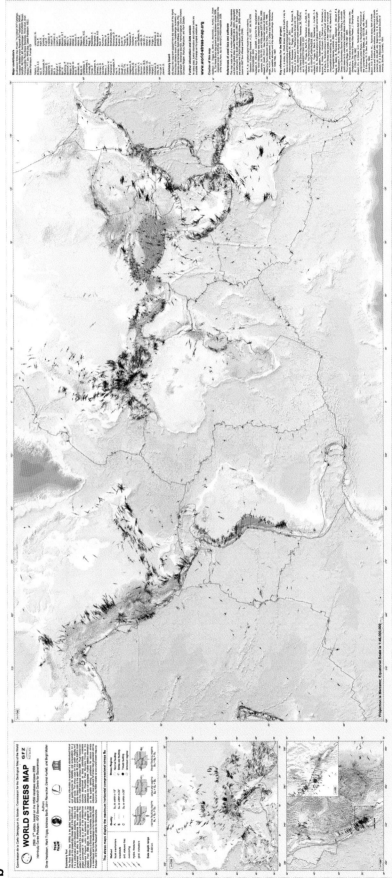

Fig. 4.3 (a) Google Earth application of the World Stress Map of Japan as an example (Image credit: ©Google earth 2012). (b) The World Stress Map is a global compilation of information on the present day stress field in the upper 40 km of the Earth's crust with 21,750 stress data records in the current WSM database release in 2008. It is a collaborative project between academia, industry and government that aims to characterize the stress patterns and to understand the stress sources (Source: Heidbach et al. 2008; Image credit: ©Google earth 2012). You can see clearly how tectonic stress is associated with plate boundaries

4.1 Folds

Folding is often demonstrated spectacularly in layered rocks, when sedimentary rocks – initially deposited as sediments in a planar structure – have been bent by either horizontally or vertically directed forces into curved, sweeping folds (Figs. 4.4, 4.5, 4.6, 4.7, 4.8, 4.9, 4.10, 4.11, 4.12, 4.13, and 4.14). The sedimentary sequences can be composed of clastic sediments (mudstone, siltstone, sandstone or conglomerates), of biological material (foraminifera, shell, coral), or less commonly chemical sediments. Folding of rock formations can occur if rocks behave in a ductile and not a brittle fashion under deformation, for example, under higher pressure or higher temperature conditions or if the deposits are young and only weakly lithified. Not only do different rock types respond in different ways but also the same rock type can show various responses depending on the conditions under which the force is applied.

There are many ways of measuring folds so that we can quantify and classify their shape, orientation, wavelength and amplitude. They can be symmetric or asymmetric about their axial surfaces or their hinge lines can be straight or curved. Another way to look at folds is to consider folds as waveforms, which can be harmonic or disharmonic through a sequence of layers. Spectacular large-scale folds in a landscape are visible when erosional processes have uncovered them or when folds are bent upward into structural arches and troughs, termed anticlines and synclines (e.g., Figs. 4.13 and 4.14). Geologists can recognize an anticline by a sequence of sedimentary beds dipping or tilting away from the center of the fold with the youngest strata located outside of the fold; whereas for a syncline, the beds will dip toward the center with the oldest beds on the outside of the fold. Folds as anticlines and synclines give vast landscapes on all continents their characteristic face (Figs. 4.4, 4.5, 4.6, 4.7, 4.8, 4.9, 4.10, 4.11, 4.12, 4.13, and 4.14).

Fig. 4.4 (**a**) At *18°03′S* and *126°28′E*, these folded and faulted sedimentary rocks are located in the northern part of Western Australia. (**b**) Steeply dipping layers of sedimentary rocks which are intensely fractured and offset along numerous fault lines in southwest Pakistan at *25°30′55″N* and *62°29′21″E*. Scene is ~6.5 km wide (Image credit: ©Google earth 2012)

b

Fig. 4.4 (continued)

Rarely are sedimentary rocks lying perfectly flat. Usually tectonic forces have created some dip, or inclination, to the horizontal. The resultant landform are cuestas, asymmetric ridges, steep on one side and gentle on the other, that form by the erosion of gently dipping rock beds (Figs. 4.15, 4.16, 4.17, 4.18, and 4.19). If the inclination of the rock formation is about 6–12°, a series of cuestas will develop because several sequences are exposed by back-cutting from the lower levels. From about 20–25° inclination of the rock sequences, more linear and vertical ridges are generally observed in the landscape.

4.2 Faults

Tectonic forces can also cause a rock formation to break and slip on both sides of the fracture – this we call a fault (Figs. 4.20, 4.21, 4.22, and 4.23). Faults come in all sizes from the grand structure of the San Andreas Fault in California (see Fig. 1.6) that can be traced for hundreds of kilometers to small faults, where the offset accounts for only a couple of centimeters. A fault can be recognized in the field by the existence of a scarp or a cliff that marks where the fault intersects with the land surface or by differences in the type and age of rock formation on both sides of the offset. There are four different major types of faults that are differentiated by the relative position of the fault plane – the flat surface along which a slip occurs during an earthquake. Normal faults occur where the crust is being pulled apart, e.g. at a divergent plate boundary, and have a fault plane that is usually dipping between 40° and 60°. The fault plane in a reverse fault is generally dipping in similar angles, but the hanging wall pushes up, and the footwall pushes down. This is the case for example at a convergent plate boundary where a plate is being compressed. Thrust faults are also caused by compressional tectonic forces and move the same way as reverse faults, but usually at an angle of ~40° or less. In these faults, the hanging wall rocks are actually pushed up on top of the footwall rocks. In a strike-slip fault the fault surface is close to vertical, and the rock formations move in opposite

Fig. 4.5 Nearly 10 km wide scene from northwest Algeria at *31°55′N* and *1°48′W* with sharp knickpoints in the folded rocks (Image credit: ©Google earth 2012)

Fig. 4.6 The Appalachian Mountains, often called the Appalachians, are the oldest mountain chain in North America and extend along the eastern margin of the continent for over 2,000 km from Newfoundland, Canada to Alabama, USA. In the valley and ridge province, Paleozoic marine sedimentary rocks are thrusted and folded into large anticlines and synclines during a mountain building episode (the Appalachian orogeny) as the two former landmasses, Gondwana and Euramerica, collided approximately 325–260 million years ago. This collision formed the supercontinent Pangaea, which involved all ancient major continental land masses. Deciduous forests composed mostly of maple, oak and hickory cover most of the Appalachian Mountains at the higher altitudes giving the landscape the scenic beauty when they are changing colors in the fall. Streams, lakes and many state parks and excellent hiking trails provide lots of recreational activities and support a flourishing tourism industry (Image credit: ©Google earth 2012)

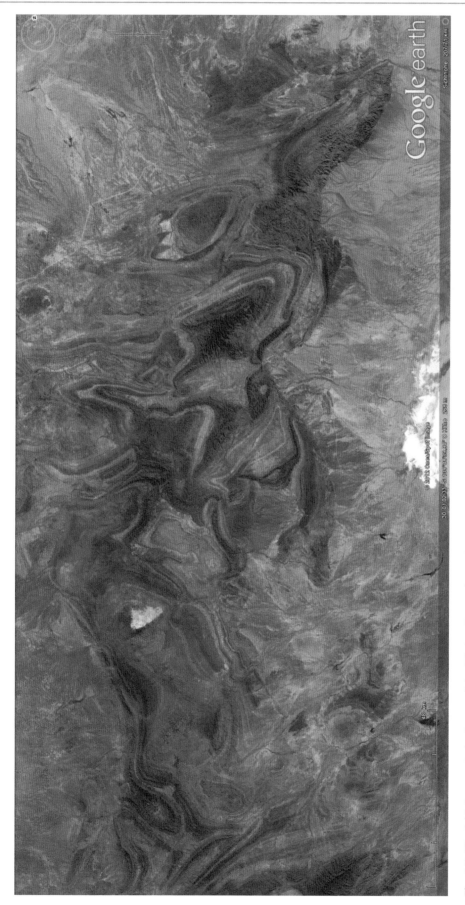

Fig. 4.7 The folded mountain landscape of the Flinders Ranges in southern Australia. The Flinders Ranges are a Y-shaped region of uplifted Neoproterozoic to Paleozoic sedimentary rocks north of Adelaide (South Australia) and attest to the influence of regional horizontal compression across the Australian Plate from the far eastern margin in New Zealand. They are home to the late Precambrian "Ediacara biota" – a diverse group of soft-bodied organisms that lived in the World's oceans about 580 million years ago. Their fossilized imprints are preserved mostly on the undersides of slabs of quartzite and sandstone (Image credit: ©Google earth 2012)

a

Fig. 4.8 (**a**, **b**) Sweeping folding structures north of Morocco's Anti-Atlas. The Anti-Atlas range formed in the Palaeozoic era during the same orogeny as the Appalachians. (**c**) An enlarged section of (**b**) (Image credit: ©Google earth 2012)

b

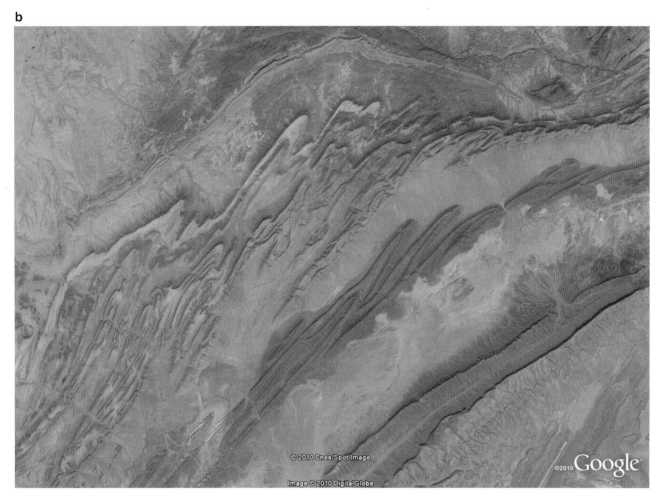

Fig. 4.8 (continued)

4.2 Faults

c

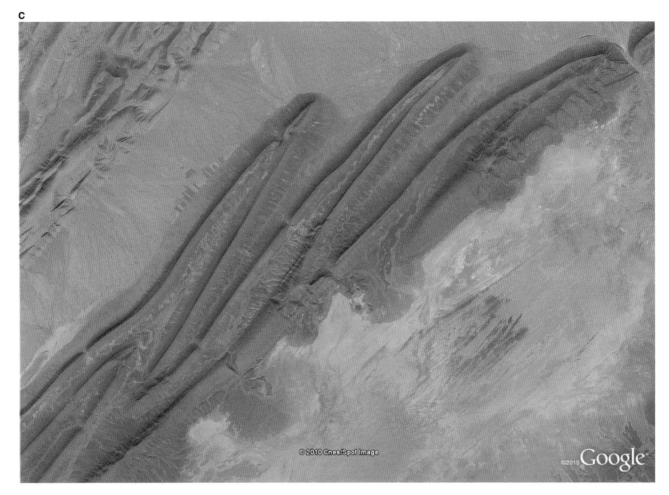

Fig. 4.8 (continued)

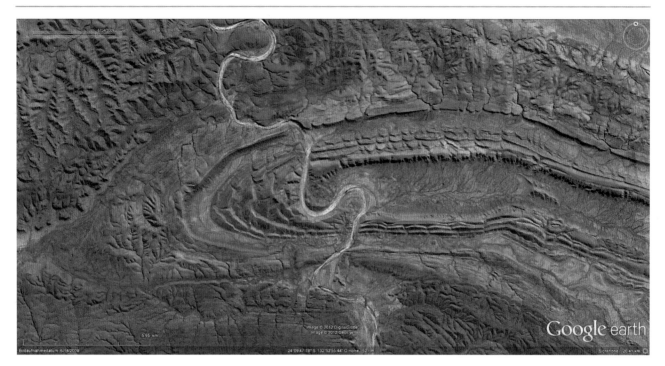

Fig. 4.9 The Finke River in central Australia crosses an anticline structure of 45 km width at *24°09′S* and *132°52′E*. Rock strata abruptly dip towards the north and south, and the anticline hinge plunges towards the west in its western part; the character of an anticline with the oldest strata in the center is evident. The valley of the Finke River is an incredibly old river system that follows exactly the same course for the past 15–20 million years and generally flowed along the same path for 100 million years. It has existed before the anticline structure was exhumed by uplifting and denudation, but the river kept its course and incised this valley in pace with the uplift. This keeping pace is called "antecedence" because the river is older than the structure it is cutting. Aboriginal mythology believes that the Finke River was formed when the Rainbow Serpent thrust north from Lake Eyre (Image credit: ©Google earth 2012)

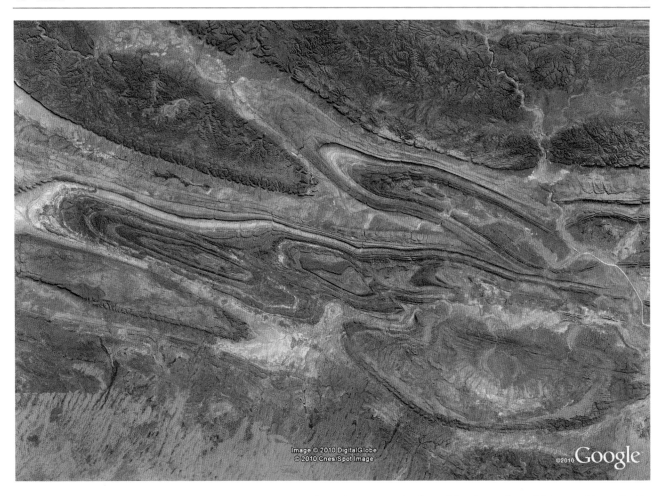

Fig. 4.10 Complex deformation patterns in the vast and spectacular sedimentary rocks of the MacDonnell Ranges near Alice Springs in central Australia. Folds are disturbed by faulting, as a result of crustal compression. Viewed from above, they speak eloquently of the caterpillar dreaming that is at the heart of local Aboriginal mythology of the Arrernte people, the traditional custodians of this area of Australia. For them, this is where ancestral creation beings – including caterpillars and wild dogs – travelled the land, naming and shaping prominent elements in the vast landscape. The scene is about 50 km wide centered at about *24°22′S* and *132°07′E* (Image credit: ©Google earth 2012)

Fig. 4.11 Overview of the detailed pictures shown in Figs. 4.9 and 4.10. The compressional folding structures are easy to follow in this image, 112 km wide and centered near *24°12′S* and *132°43′E*. The Gosse's Bluff impact crater (see Chap. 5) is visible in the northern part of the image (Image credit: ©Google earth 2012)

a

Fig. 4.12 (**a**) One of the most spectacular examples of anticlinal fold structures in the world can be seen along the shores of the Persian Gulf, on the north shore of the Strait of Hormuz. These anticlines form the foothills of the Zagros Mountains, which run north-northwesterly through Iran, and were formed when the Arabian shield collided with the western Asian continental mass some 5–10 million years ago. The compression and shortening expressed by the folds are accompanied by extensive thrust faults. All the deformation is rather young; the folded sediments are of Paleogene and Neogene age. Anticlinal structures are well known as classic traps for hydrocarbons and some producing wells are located in the area seen in the image. The dark circular patches represent the surface expression of salt domes (see also Figs. 4.36 and 4.37) that have buoyantly risen from the Cambrian Hormuz salt horizon through the younger sediments to reach the surface. Only in a hot arid environment such as Iran the soluble salt can escape rapid erosion. (**b**) In northwestern Saudi Arabia (opposite to the Sinai Peninsula) at about *28°18′N* and *34°50′E*, the slightly deformed younger and light colored sedimentary rocks (syncline, mostly sandstones) form a basin structure and cover the dark and older basement rocks (Image credit: ©Google earth 2012)

4.2 Faults

b

Fig. 4.12 (continued)

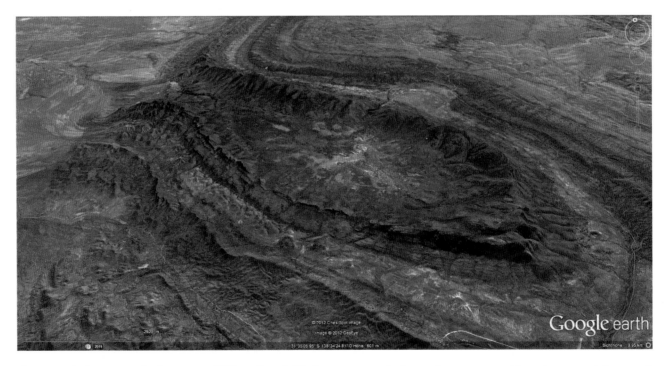

Fig. 4.13 The famous synclinal structure of Wilpena Pound, Flinders Ranges (South Australia) forms an almost closed basin. Mountain ranges surrounding the remnant valley have been eroded away during the last several million years. The higher walls of Wilpena Pound are composed of quartzite, a metamorphic rock converted from sandstone by elevated temperature and pressure, which is very resistant to weathering (Image credit: ©Google earth 2012)

Fig. 4.14 A beautiful syncline in the Andes Mountains of northwest Argentina, exposing Jurassic rock formations. Large-scale erosion accompanying the uplift during the Andean orogeny has removed the adjacent (anticline) parts of these rock formations, but the trough-like syncline persisted. Picture-perfect cuestas have developed towards the center of the structure, where inclined layers of sedimentary rocks of different colors are incised by draining channels (Image credit: ©Google earth 2012)

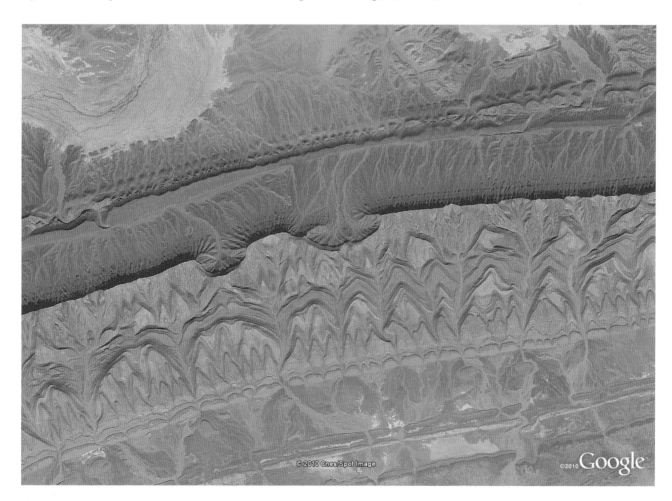

Fig. 4.15 If the inclination of rock layers is significant (here with about 8° to the south in southern Morocco), the cuestas show a differentiated dissection by infrequent, but severe rainfalls (*28°19′N, 9°44′W, 38 km wide*) (Image credit: ©Google earth 2012)

Fig. 4.16 An asymmetrical anticline with cuesta formations exposing older rock in the "geologic window" of the center. The site is at *26°56′N* and *54°01′E* in Iran and the scene is more than 30 km wide (Image credit: ©Google earth 2012)

Fig. 4.17 (**a**) Long, parallel and straight cuestas built by steeply dipping strata in the northern part of Valley of Fire National Monument, Nevada (USA), (**b**) A similar aspect in the Flinders Ranges of southern Australia (Image credit: ©Google earth 2012)

4.2 Faults

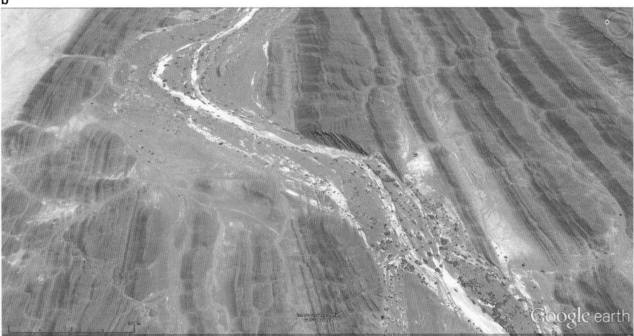

Fig. 4.18 (**a**) Typical landscape aspect of a long and straight cuesta. The inclination of rock layers is clearly visible in the Flinders Ranges (South Australia) (Image credit: S.M. May). (**b**) Close-up of cuestas in Pakistan from Google Earth. The typical inclination of the formerly flat-lying rock layers is visible from space (Image credit: ©Google earth 2012)

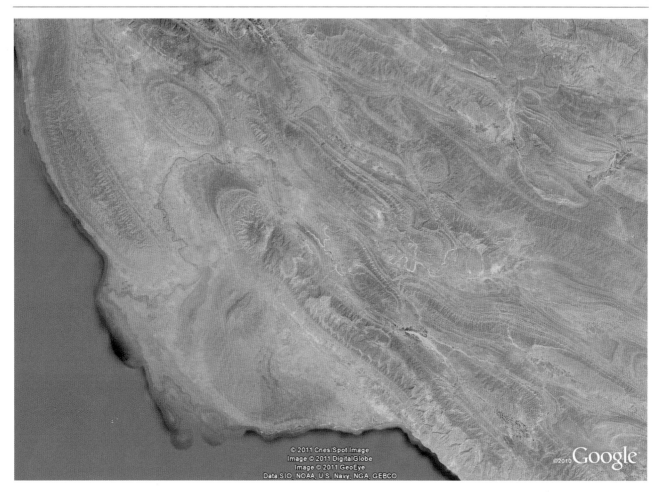

Fig. 4.19 Western Iran is a landscape with elongated anticlines pushed upwards by salt intrusions, forming a vast landscape of cuestas (Image credit: ©Google earth 2012)

Fig. 4.20 A fault line in southern Israel near the Aqaba graben structure at *29°36′N* and *34°54′E*. Width of scene is 8.6 km (Image credit: ©Google earth 2012)

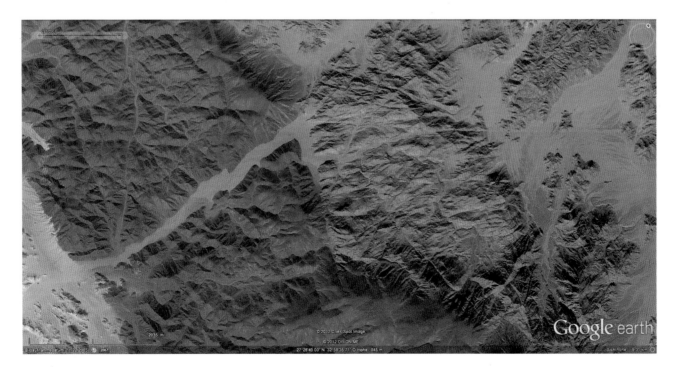

Fig. 4.21 Fault line in northern Sudan at around *27°28′N* and *32°58′E* (Image credit: ©Google earth 2012)

Fig. 4.22 (a) The Salar de Navidad fault in N Chile (stretching from west-northwest to east-southeast) is crossed by several drainage channels, resulting in a contrasting hydrological drainage pattern on both sides of the fault. Slight lateral offset of channels and sediment deposition to the south of the fault line are visible. (b) The Paposo normal fault (i.e. a fault with predominant vertical motion) in N Chile. Along the fault scarp, numerous alluvial fans have formed due to vertical movement and erosion of the W part (Allmendinger and González 2010) (Image credit: ©Google earth 2012)

4.2 Faults

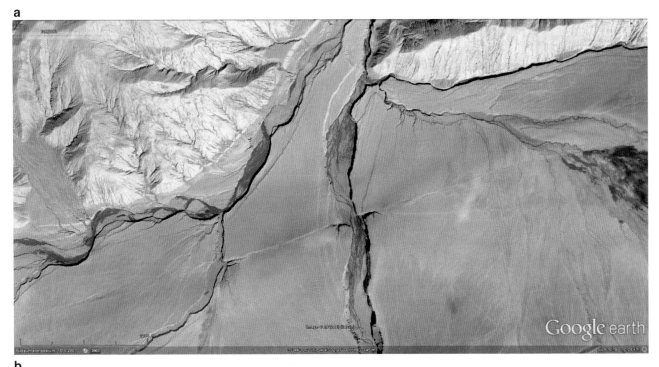

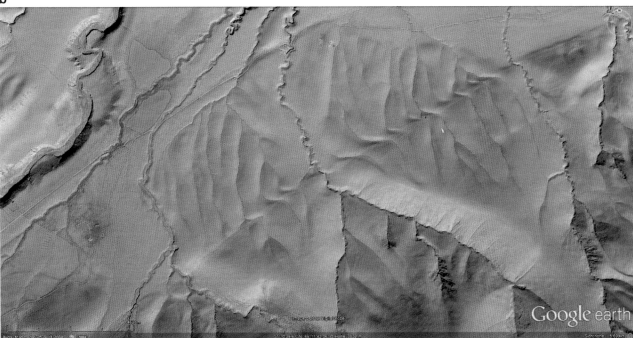

Fig. 4.23 (**a**) A section of the Kunlun strike-slip fault (i.e., a fault with predominant lateral motion) in Central China. The Kunlun fault is one of the major faults of the Indo-Asian collision zone. The fault line crosses the image from west-southwest to east-northeast. Lateral displacement along the fault line can be inferred in this image from discontinuities of the channel course, differences in fluvial terrace formation and lateral offset of terrace edges and channels. (**b**) Section of the Altyn Tagh strike-slip fault zone which is another major fault related to the Indo-Asian collision zone, forming the northwestern boundary between the Tibetan Plateau and the Tarim Basin. Lateral offset of smaller fluvial channels is clearly visible (**c**) A different section of the Altyn Tagh fault zone, appearing as a straight line in the image. While river channels show incision south of the line, sediment is accumulated in extensive alluvial fan systems directly to the north. (**d**) In northern Iran in the southern part of the Semnan province, geologically young (i.e., only several million years old) basin sediments have been deformed by tectonic forces including several parallel faults of more than 130 km in length, striking west-southwest to east-northeast. The figure shows a section of 9 km (Image credit: ©Google earth 2012)

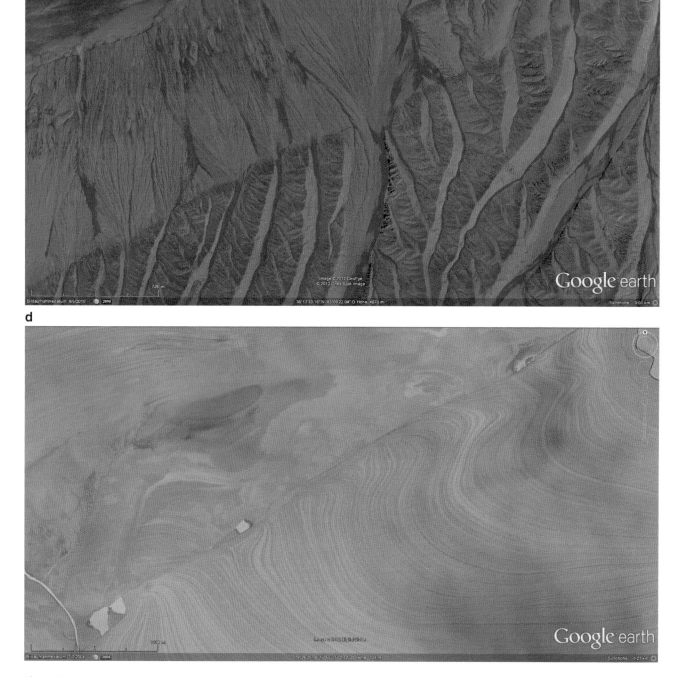

Fig. 4.23 (continued)

4.3 Joints

A joint is a type of rock fracture where no movement on either side can be observed – you may call it simply a crack in the rock (Figs. 4.24, 4.25, 4.26, 4.27, 4.28, 4.29, 4.30, 4.31, 4.32, and 4.33). Joints can be caused by tectonic stress, but also may form when the rock is exposed to differences in temperature regimes and reacts with contraction and expansion such as during the cooling of lava or frost weathering in cold climates. Water, air and other erosional agents will exploit these cracks, accelerate weathering and erosional processes, and weaken the internal structure of the rock formation. The circulation of hydrothermal solutions through joints can deposit minerals such as calcite or quartz that form veins in their host formation.

4.4 Circular Structures

As we have seen, deformation along plate boundaries by horizontally directed forces tend to produce planar faults or folds orientated parallel to the plate boundary. Other tectonic forces cause deformation structures that are usually more symmetrical and form circular structures that tend to be more common in the interior of plates, far away from active plate boundaries. Different types of deformation can produce circular geomorphologic structures that are termed basins and

Fig. 4.24 A joint pattern in a limestone landscape in southern Andalusia, Spain, at *36°57′N, 4°33′W*. Scene is 3.7 km wide (Image credit: ©Google earth 2012)

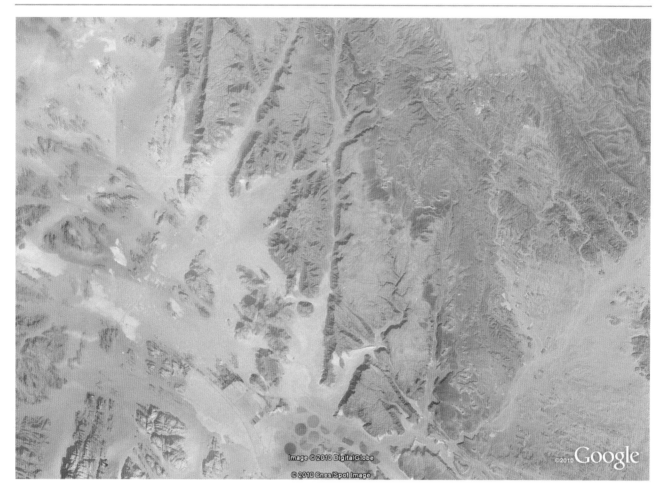

Fig. 4.25 Long dominant joints in direction of the Aqaba graben in southern Jordan at *29°41′N* and *35°36′E* in a 61 km wide section (Image credit: ©Google earth 2012)

4.4 Circular Structures

Fig. 4.26 The rock formation in southern Jordan (*29°37′1.29′N, 35°29′54.34′E*) has been deeply incised by erosion along the main joints – more and more small jointing patterns are exposed by ongoing weathering and subsequent erosion during infrequent but heavy rains (6 km wide scene) (Image credit: ©Google earth 2012)

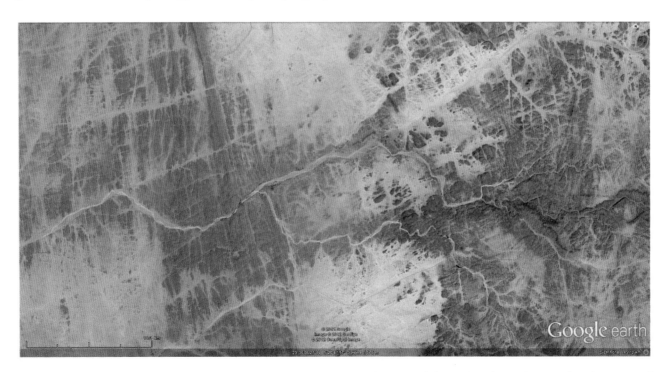

Fig. 4.27 Intersecting joints exposed by weathering in southern Algeria at *23°44′N* and *4°59′E* in a 30 km wide section. Note how the courses of the rivers follow the joint pattern (Image credit: ©Google earth 2012)

Fig. 4.28 (a) Intensely weathered joints in very old rocks (about 2 billion years) in northwest Australia at *12°50′S, 133°23′E* (10 km wide section). There is a "hierarchy" of dominant primary and secondary joint patterns, which likely is either the result of different tectonic stress or just an expression of time of exposure to weathering processes to explore even the tiniest weak fractures in the rock. (**b**) Detail of the landscape in Fig. 4.8a, 5 km wide (Image credit: ©Google earth 2012)

4.4 Circular Structures

Fig. 4.28 (continued)

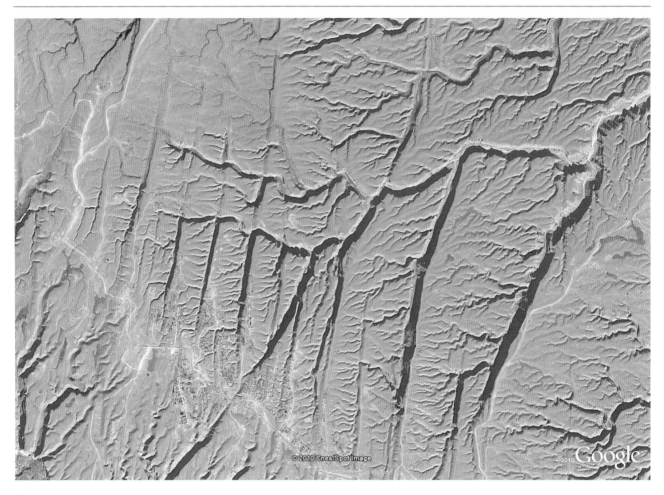

Fig. 4.29 Structures (a joint pattern) controlling a valley system in central Pakistan at *29°09′N* and *69°06′E*. Width of scene is 18 km (Image credit: ©Google earth 2012)

4.4 Circular Structures

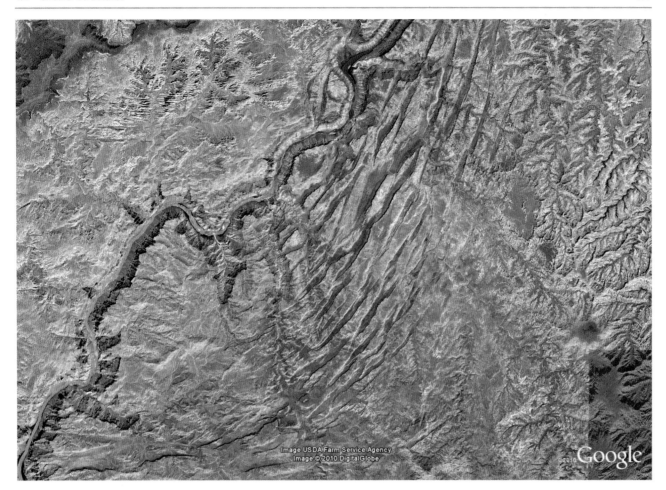

Fig. 4.30 In Canyonlands National Park of Utah, USA (*38°06′N* and *109°57′W*), an invisible salt dome under the surface spreads the rocks at the surface, resulting in curved, parallel joint patterns. Width of scene is 48 km (Image credit: ©Google earth 2012)

Fig. 4.31 (**a, b**) The Arches National Park in Utah, USA, exhibits fine examples of jointing caused by an anticline uplift of a salt dome which has stretched and pulled apart the overlying rock strata (*38°47′N, 109°36′W*; about 4.5 km wide scene). (**c**) Detailed image showing secondary joints cross-cutting the dominant joints. Width of this image is just 2 km (Image credit: ©Google earth 2012)

4.4 Circular Structures

b

Fig. 4.31 (continued)

c

Fig. 4.31 (continued)

4.4 Circular Structures

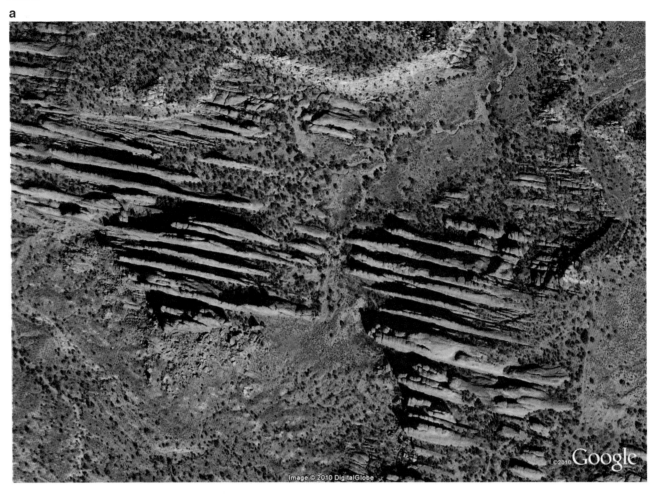

Fig. 4.32 (a) In Arches National Park (Utah, USA), main joints have been weathered to more than 30 m deep leaving behind delicate, fin-like structures. Scene is 3 km wide at *38°50′N* and *109°41′W* (Image credit: ©Google earth 2012). (b) Rock fins in Arches National Park (Utah, USA). The aspect shows a landscape as in the satellite image of (a) (Image credit: D. Kelletat)

b

Fig. 4.32 (continued)

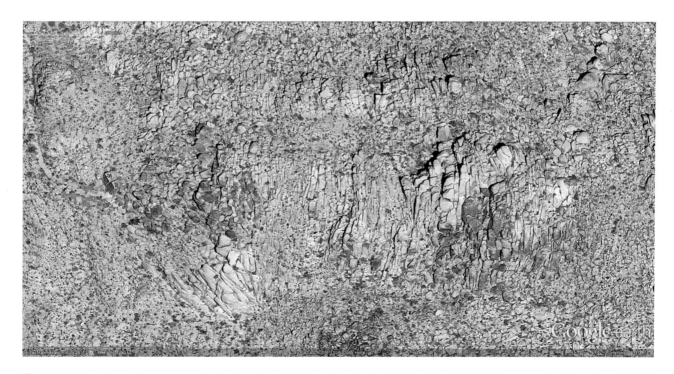

Fig. 4.33 Joint-orientated corestone weathering (see Chap. 6) of granitic rocks in the Joshua Tree NP, USA (Image credit: ©Google earth 2012)

4.4 Circular Structures

domes – the process of basin formation can vary. Deformation can be caused by rising bodies of buoyant material such as magma or salt (salt tectonics) that push overlying sediments upwards to form domes or by downward forces where materials sinks or subsides. Some sedimentary basins form when a heated portion of the lithosphere cools and contracts and cause the overlying rock or sediment formations to subside (thermal subsidence basins). Other types of basins form when tectonic forces stretch and thin the Earth's crust (rift basins) or compress the crust downwards (flexural basins). Also the weight of sedimentary sequences in large river deltas can cause the crust to subside and form a sedimentary basin.

A dome is a broad circular or elliptical upward bulge of rock layers with the flanking beds dipping away in all directions from a central point. They often form spectacular scenic landscape features that extend many kilometers in diameter such as Upheaval Dome in southeast Utah, USA (Fig. 4.34) or the famous Richat Structure in Mauretania (Fig. 4.35). Some domes even extend for hundreds of kilometers. Since gypsum and rock salt are easily dissolved by water (rain, meltwater or even dew), salt tectonics and associated landforms such as salt domes can only develop and exist for a longer time in arid regions like hot and very cold deserts (Figs. 4.36 and 4.37).

A basin is a bowl-shaped depression of rock layers in which the layers of rock dip towards a central point. Basins are excellent sediment traps and in some cases, deposition of sediments can accumulate sedimentary sequences many kilometers in thickness. Different coloring of the sediments can enhance this visual impression: Dark and blue colored sediments may stem from lake or lagoon sediments; whereas red, orange and yellow colors mostly occur after the sediments have been exposed to weathering processes at the atmosphere-rock interface and iron, aluminum, manganese and other minerals are enriched at the outer surface of individual grains (Fig. 4.38). Where erosion by wind and water exposes the interior stratification of these basin sediments three-dimensional natural artworks of astonishing beauty and aesthetics are created. Good examples are shown in Figs. 4.39 and 4.40. The series of images show examples of basins from different continents with an emphasis on arid environments uncovered by soil and vegetation.

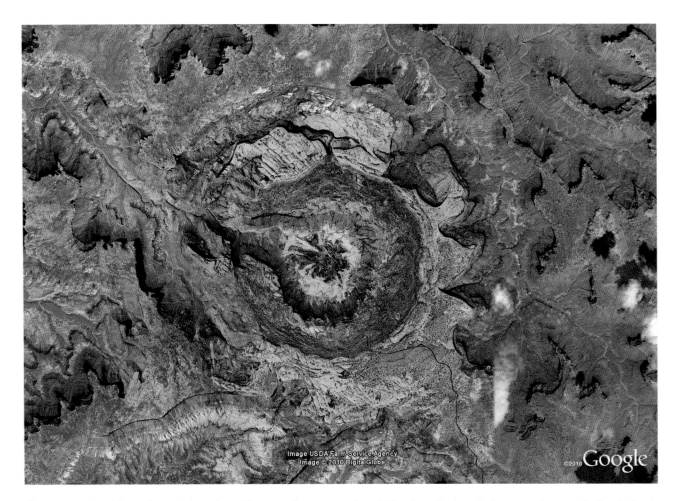

Fig. 4.34 Upheaval Dome is a salt intrusion with a cap of gypsum pushing up the top sandstone layers in Canyonlands National Park, Utah, USA, at *38°26′N* and *109°55′W*. The round structure has a diameter of about 3.2 km. In earlier days, the feature was interpreted to be of volcanic origin or as an impact crater (Image credit: ©Google earth 2012)

Fig. 4.35 The famous Richat Structure. This prominent circular feature in the Sahara Desert of Mauritania (northwest Africa) has attracted much attention since the earliest space missions because it forms a conspicuous bull's-eye that has become a landmark for shuttle crews in the otherwise rather featureless expanse of the desert. Initially interpreted as a meteorite impact structure because of its high degree of circularity, it is now thought to be merely a symmetrical uplift that has been sculptured bare by erosion. Palaeozoic quartzites form the resistant beds outlining the structure (Image credit: ©Google earth 2012)

Fig. 4.36 The collision between the Asian landmass and the Arabian platform has folded rocks and pushed up the rugged Zagros Mountains in southern Iran. In places, underlying deposits of salt have ascended in viscoplastic plumes and pushed through the top layers of sedimentary rocks because of lower density and plasticity of the salt. Due to the solubility of halite or gypsum these landforms can only exist in very arid climates (deserts), or in cold latitudes, where all water is frozen. Similar to limestone areas, karst-type landforms may form due to solution of the salt. In humid areas salt deposits will be dissolved by groundwater several 100 m below the surface. Once the salt extrudes at the surface, gravity causes the salt to flow like glaciers into adjacent valleys. The resulting tongue-shaped bodies can extend for several kilometers and usually are sculptured with repeating bow-shaped ridges separated by crevasse-like gullies. The darker tones visible in the images are due to clays brought up with the salt, as well as the probable accumulation of airborne dust (Image credit: ©Google earth 2012)

4.4 Circular Structures

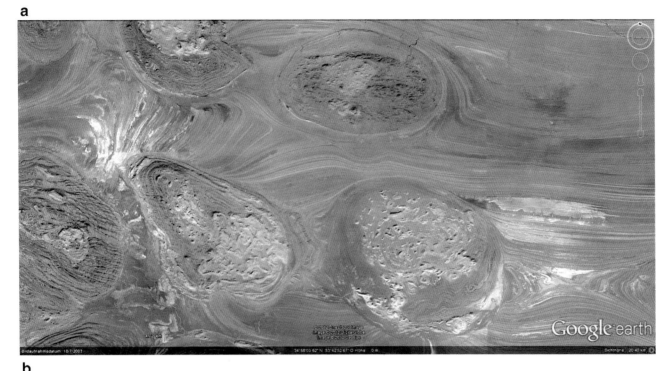

Fig. 4.37 (**a**) Salt domes in Iran. Salt domes are also common trap sites for hydrocarbons. (**b**) Salt dome with salt exposed at the surface in the inner part of the structure. The rock strata around the center have been folded and faulted due to deformation related to the rising salt dome below (Image credit: ©Google earth 2012)

a

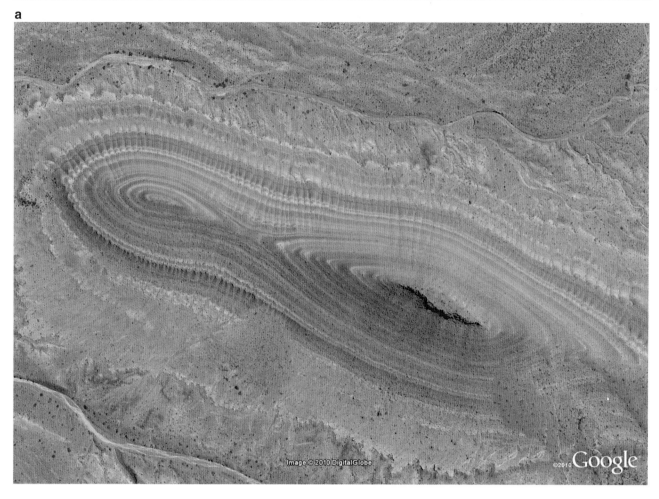

Fig. 4.38 (**a**, **b**) In central Patagonia (Argentina) at *37°51′S* and *69°44′W*, layers of rock strata in a mudstone/siltstone sequence can be distinguished by their changing colors (Image credit: ©Google earth 2012)

4.4 Circular Structures

Fig. 4.38 (continued)

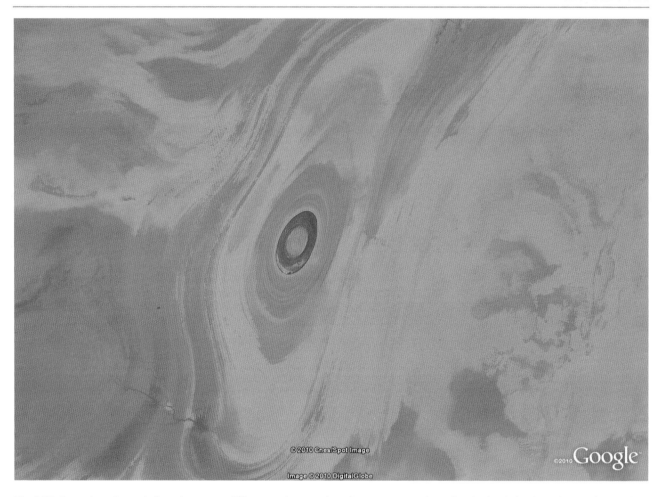

Fig. 4.39 In northern Iran, wind erosion expose different mudstone and sandstone sequences in ancient basins (related to salt tectonics) at about *34°42′N* and *55°11′E*. Width of scene is 12 km (Image credit: ©Google earth 2012)

4.4 Circular Structures

a

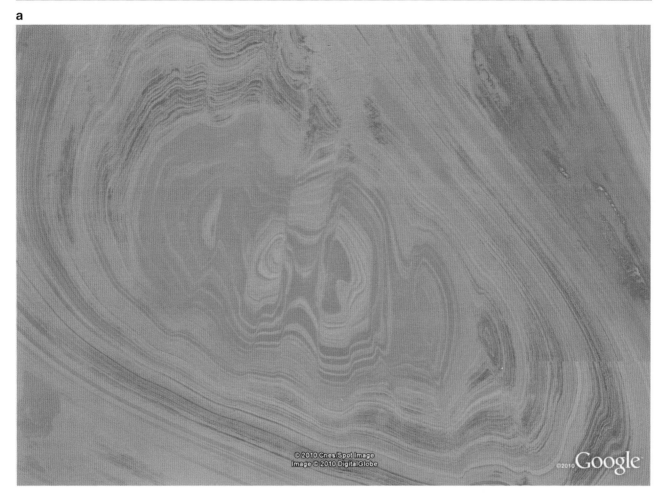

Fig. 4.40 (**a**, **b**) Deformation of exposed silt and clay layers in northern Iran at *34°38′N* and *54°01′E* and *34°51′N* and *53°47′E*. Width of scene is approximately 6 km (Image credit: ©Google earth 2012)

b

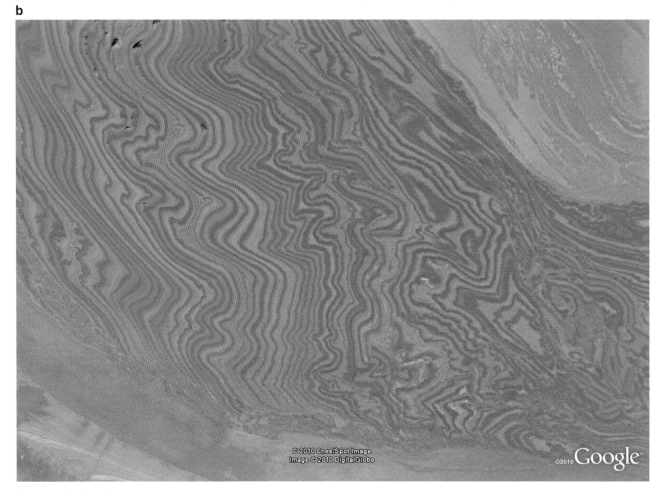

Fig. 4.40 (continued)

Further Readings

Allmendinger RW, González G (2010) Neogene to quaternary tectonics of the coastal cordillera, northern Chile. Tectonophysics 495:93–110

Collier M (1999) A land in motion: California's San Andreas fault. University of California Press, Berkeley

Fossen H (2010) Structural geology. Cambridge University Press, Cambridge

Heidbach O, Tingay M, Barth A, Reinecker J, Kurfeß, D, Müller B (2008) The World Stress Map database release 2008. doi:10.1594/GFZ.WSM.Rel2008

Kirby E, Harkins N, Wang E, Shi X, Fan C, Burbank D (2007) Slip rate gradients along the eastern Kunlun fault. Tectonics 26, TC2010. doi:10.1029/2006TC002033

McKnight TM, Hess D (2000) The internal processes: types of faults. Physical geography: a landscape appreciation. Prentice Hall, Upper Saddle River, pp 416–417

Pavitt N (2001) Africa's great rift valley. Harry N. Abrams, New York

Part III

Exogenic Forms and Processes

Impacts Craters

Abstract

In 1980, father and son geologists Luis and Walter Alvarez discovered unusually high traces of iridium in rock layers near Gubbio (Italy) that define the Cretaceous-Paleogene boundary (K-Pg; formerly known as the K-T boundary) when about 75 % of all species on Earth, including the dinosaurs, perished forever in a mass extinction event. The Alvarez's put forward an extraordinary claim – that the accumulation of this much iridium-bearing dust would require an asteroid of about 10 km wide to hit the Earth. The search was on. But it proved to be difficult for two reasons: First, the event took place 65 million years ago and the crater could have been eroded or filled with sediments. And secondly, most of the Earth's surface is covered with water making it difficult to detect craters on the ocean floor. A decade later, geologists and geophysicists searching for oil finally found a huge crater that fit the search criteria. It was almost 200 km in diameter and 1.5 km deep, but buried under sediments near a town called Chicxulub on Mexico's Yucatán Peninsula. *Chicxulub* means "tail of the devil" in the local Mayan language. Today, almost all scientists agree that this gigantic asteroid impact produced a blast 6 million times more powerful than the 1980 eruption of Mt. St. Helens and caused a cataclysmic collapse of ecosystems that lead ultimately to the K-Pg mass extinction. The last piece of the puzzle was put together in 2013, when researchers provided new evidence that the age of the Chicxulub asteroid impact and the K-Pg boundary coincide precisely. This short chapter presents a few visual examples of impact craters that are visible on the face of the Earth and explains why they are so scarce and rarely preserved on Earth.

Most of the known asteroids (termed so by William Herschel, meaning star-like) within our Solar System orbit in the asteroid belt located between Mars and Jupiter. Nobody knows for certain how many objects occur in this belt, but several hundred thousand asteroids have been discovered and given provisional designations so far. Thousands more are discovered each year. Although these numbers are staggering indeed, the asteroid belt stretches out over a vast region and is so thinly populated that several unmanned spacecraft have been able to move through it – this isn't the dense field of asteroids our intergalactic heroes have to dodge as you see in movies like *Star Wars*!

Over 200 are known to be larger than 100 km in diameter and at least a few million are larger than 1 km. Most of them orbit the Sun without any threat to the Earth, but some have more risky orbits that bring them on occasional visits into the inner solar system and near enough to Earth to have reasons for concern. These asteroids are called Near-Earth Asteroids (or simply Near Earth Objects, NEO's) and astronomers have catalogued over 7,000 so far with nearly 1,000 of these objects bigger than 1 km.

Impact craters are evidence that sometimes asteroids are on collision course with our planetary home and recently in 2013 the nation of Russia was caught by surprise when a spectacular cosmic coincidence on February 15th 2013 provided a vivid reminder that the solar system can sometimes turn into a shooting gallery: A meteor (meteorites are smaller fragments of an asteroid) with a diameter of many meters

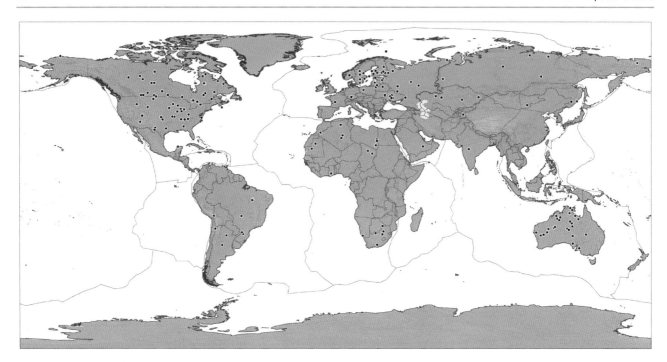

Fig. 5.1 Confirmed impact craters on the continental landmasses of the Earth based on the Earth Impact Database (URL: http://www.passc.net/EarthImpactDatabase/)

entered the atmosphere with a speed of 18 km/s (64.800 km/h) and disintegrated in the skies over Tschelyabinsk, Russia. The energy released by the explosion in the lower atmosphere with a light flash was in the hundreds of kilotons TNT. The explosive shock damaged buildings in six cities and injured more than 1,000 people, and left behind an impact signature on the frozen Lake Chebarkul.

Later in 2013, the largest remnant of this meteorite has been excavated from a small pond. It consists of nickel and iron and is more than 1 m in diameter and about 600 kg. The trajectory of the Russian meteor was significantly different than the trajectory of the asteroid 2012 DA14, which hours later made a close but safe flyby of Earth (well between the orbit of the moon), making it a completely unrelated, but spectacular coinciding cosmic event. The Russian meteor is the largest reported since 1908 AD, when a meteor or comet hit Tunguska, Siberia.

The Tunguska meteorite (or comet?) exploded above the surface by friction in the atmosphere. The shockwave pushed down all trees on several 100 km^2 of land and the hot fragments set many of them alight. An explosion in the atmosphere seems to be characteristic of small objects entering the atmosphere. The Tunguska object is estimated to have a size of several meters and some thousands of tons and nearly all was transformed into energy by its explosion many kilometers high in the atmosphere.

As of today we know of nearly 200 impact craters on Earth of very different size and age (Earth Impact Database 2010; see also Fig. 5.1). This is in stark contrast to the surface of our Moon, which is literally peppered with thousands of well-preserved impact craters of every size from the time of the Late Heavy Bombardment between 4.1 and 3.8 billion years ago, an incredibly violent period in the history of our Solar System. The Late Heavy Bombardment also showered the Earth and other inner planets with cosmic debris. But what triggered this cosmic cataclysm? Cosmologists are able to simulate the early Solar System in great detail and it is now thought (although still controversial) that the outer gas planets Saturn, Uranus, Jupiter and also Neptune formed much closer to the Sun compared to their current position in the Solar System. The trigger was a gravitational surge because Jupiter and Saturn aligned in exactly the same position in their orbits that plunged the Solar System into an unstable era and resulted in the outwards migration of the giant outer planets. On their journey they perturbed the orbits of the asteroids and some were thrown into the inner Solar System where they collided with the Moon and the terrestrial inner planets, including Earth.

Impact scars are preserved on all terrestrial planets and the lunar surface, but in contrast to Mercury, Venus or Mars, whose surfaces are essentially frozen in time, our home planet in the universe preserves very few geologic memories of the violent ancient bombardment. Plate tectonic processes have almost completely recycled and renewed the surface of our planet during the last billions of years. Weathering and erosion resulting from the interaction of the plate tectonic systems and Earth's climate system are constantly reshaping the Earth's surface topography, destroying certain morphologic

Table 5.1 Potential effects of impacts by asteroids and meteorites

	Example or size equivalent	Most recent	Planetary effects
Supercolossal: radius $(R) > 2,000$ km	Moon-forming event	4.51×10^9 years ago	Melted part of planet
Colossal: $R > 700$ km	Pluto	More than 4.3×10^9 years ago	Melted part of crust
Huge: $R > 200$ km	4 Vesta (large asteroid)	$\sim 4 \times 10^9$ years ago	Vaporized oceans
Extra-large: $R > 70$ km	Chiron (largest active comet)	3.8×10^9 years ago	Vaporized upper 100 m of oceans
Large: $R > 30$ km	Comet Hale-Bopp	$\sim 2 \times 10^9$ years ago	Heated atmosphere and surface to $\sim 1,000$ K
Medium: $R > 10$ km	*Chixculub* event; 433 Eros (largest near-Earth asteroid)	65×10^6 years ago	Caused fires, dust, darkness, chemical changes in ocean and atmosphere; large temperature swings
Small: $R > 1$ km	Average size of near-Earth asteroids	~300,000 years ago	Caused global dusty atmosphere for months
Very small: $R > 100$ m	Tunguska event (Siberia)	1908	Knocked over trees over kilometers away; caused minor hemispheric effects, dusty atmosphere

Source: Lissauer (1999)

evidence over time. Fortunately for us, the Earth has an atmosphere that not only sustains life, but also protects us from smaller cosmic projectiles on collision course with Earth day after day. When a meteoroid enters the atmosphere with a velocity ranging from 11 to 72 km/s, it immediately begins to compress the thickening atmospheric gases, which begin to heat up. This in turn will heat up the meteoroid until it is white-hot and radiates visible light. In its next passage of journey, the meteoroid will rapidly decelerate as it encounters increasingly denser portions of the atmosphere, especially in the layers below 12 km where 90 % of Earth's atmospheric mass lies – similar to firing a bullet into water. When the meteoroid finally collides with the surface of the Earth, we call it a meteorite.

Earth accumulates some 40,000 tons of extraterrestrial material from space each year, mostly as dust and unnoticed small objects. Astronomers try to calculate the frequency of impacts for different size classes of asteroids and meteorites and came to numbers of several thousand during the history of mankind. And a cosmic chunk of matter 1–2 km in size still collides with Earth every few million years or so. We have seen that a collision with a 10-km asteroid 65 million year ago caused a mass extinction in which 75 % of all species including the largest animals ever walked on Earth, the dinosaurs, vanished from the surface of the Earth. However, these destructive impacts can also be seen as important driving forces for evolution and life on Earth as they create new opportunities and ecological niches in which new life forms and species can evolve – as the mammals and thus our ancestors and ourselves did after Chicxulub.

Table 5.1 describes the potential effects of impacts by asteroids and meteorites.

Google Earth is an excellent tool to hunt for impact craters with their distinct morphology. Simple impact craters usually form a bowl-shaped depression and have raised rims (Figs. 5.2, 5.3, 5.4 and 5.5), even if a meteorite hits the surface at a very low angle. The shockwave caused by the impact fractures the rock and excavates a large cavity depending on the mass of the meteorite, generally about 20 times larger than the impactor itself. The meteorite itself is shattered into small pieces or may melt or vaporize. More complex craters can have terraces, central peaks, and multiple rings. If the cosmic chunk is large enough, some of the material pushed toward the outer slopes of the crater will slump back toward the center, and rock beneath the crater will rebound due to the elasticity of the lithosphere and create a central peak in the crater.

Radiometric dating has revealed that five known impact craters are younger than 1,000 years and 16 are younger than 1 million years. The oldest impact has been dated to 2.4 billion years and remarkably, is still visible on the surface. Researchers studying ancient myths and their geologic and archaeological traces think that Pleistocene and Holocene impacts and other celestial phenomena are reflected in the myths of ancient cultures that have survived for hundreds and even thousands of years. Ample examples can be found in Australian Aboriginal astronomy records. The Henbury crater field in central Australia was formed when a fragmented iron-nickel meteoroid struck the central Australian desert $4,200 \pm 1,900$ years ago and excavated 13 craters covering an area of approximately 1 km^2 (Fig. 5.6). This event was probably witnessed first-hand by aboriginal people and there exists enticing evidence that the memory is passed on to modern times. Several old impact craters are shown in Figs. 5.7, 5.8 and 5.9, but the degree of preservation of their morphologic character is not always congruent with their age.

Several large and old impact craters have been preserved in the old landmass of Quebec (Canada) as this part of our world has never experienced episodes of higher sea level or extended accumulation or erosion of sediments. Nevertheless,

Fig. 5.2 The Meteor or Barringer Crater in northern Arizona (USA) is the best preserved crater on Earth with a diameter of 1.2 km and a depth of 170 m. It has been formed about 49,000 years ago when an iron-nickel meteorite with a diameter of approximately 50 m and a weight of 300,000 tons travelling at 11 km/s struck the ground. During the explosion (believed to be equal to about 10 megatons of TNT) the sandstone and limestone rock formations were bent upwards to form the outer ring. Impacts of this size statistically occur on Earth every 1,000 years (Image credit: ©Google earth 2012)

the excellent preservation of the Clearwater craters (Fig. 5.10) and of Manicouagan crater (Fig. 5.11a, b) is astonishing because during warmer climate episodes, tropical, deep chemical weathering tend to change exposed bedrock into more easily eroded regolith. During the Quaternary, the region has experienced many glacial cycles during which the movement of ice and glaciers did an immense amount of geologic work in forms of erosion, transportation and deposition and changed the topography of the landscape forever.

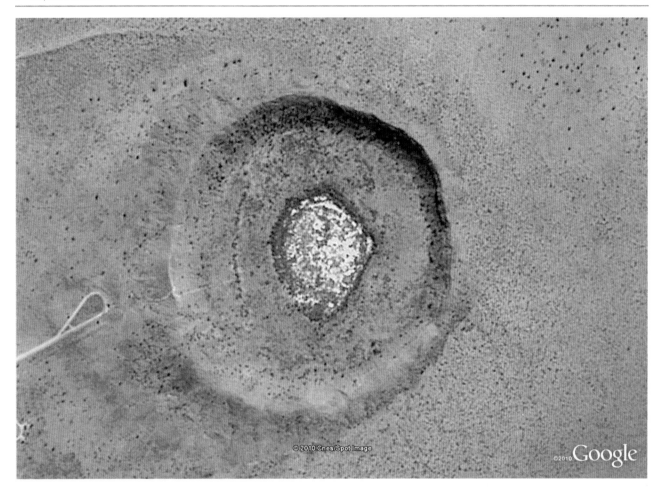

Fig. 5.3 Wolfe Creek Crater in Western Australia at about *19°10′S* and *127°47′E*. Diameter is 875 m with a depth of 60 m. The impactor is estimated to have had a weight of 50,000 tons of iron and nickel striking the ground at about 15 km/s. The age of the crater has been determined to be around 300,000 years (Image credit: ©Google earth 2012)

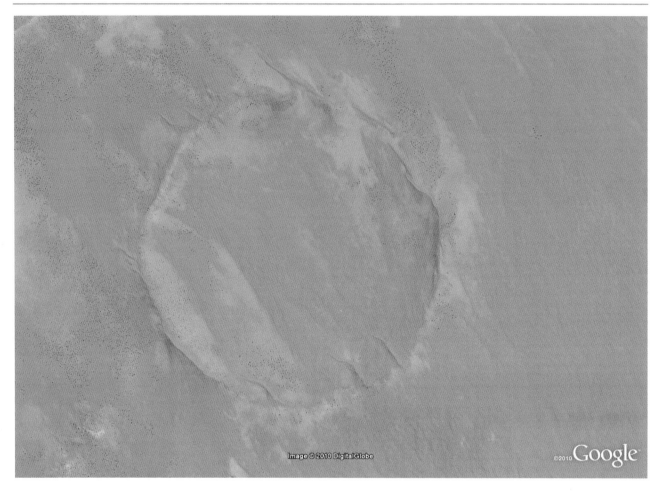

Fig. 5.4 Roter Kamm crater in Namibia at *27°45'S* and *16°17'E* with a diameter of 2.5 km, depth of 130 m and an age of 3.7 million years. It exhibits a perfect round form in very old gneiss but has been filled by shifting sand in the desert environment and will possibly be covered in the near future. Melt breccias found in the rim are strong evidence for an extraterrestrial impact (Image credit: ©Google earth 2012)

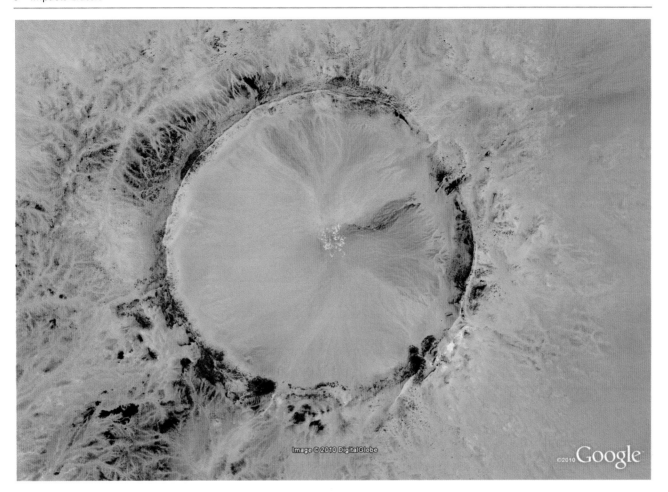

Fig. 5.5 The superbly preserved Tenoumer crater in Mauretania, northwest Africa, at about *22°53′N* and *10°24′W* has an age of only 22,000 years. The site is composed of very old Precambrian rocks. The diameter is around 1.9 km and its depth near 110 m, making the likely size of the impactor 200 m across or more (Image credit: ©Google earth 2012)

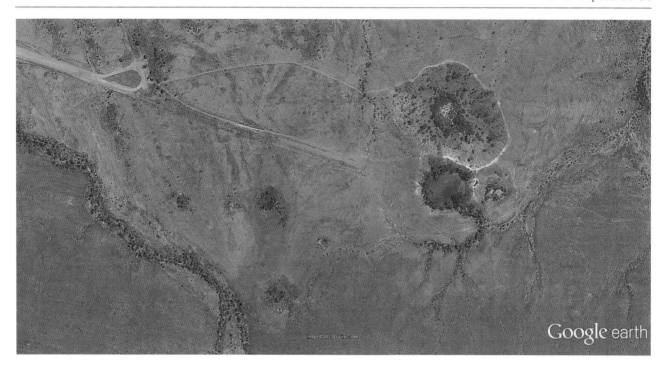

Fig. 5.6 The Henbury crater field in central Australia (Northern Territory). Aboriginal oral traditions about the crater field are sacred and secret to outsiders as well as other cultural knowledge about the physical world. The interested reader may refer to the excellent blog of Dr. Duane Hamacher (URL: http://aboriginalastronomy.blogspot.com.au/). Hamacher (2012) highlights an aboriginal oral story from the central desert regarding its cosmic origins closely parallels the scientific explanation: *"During the Dreaming, a group of sky-women were dancing as stars in the Milky Way. One of the women grew tired and placed her baby in a wooden basket, called a turna. As the women continued dancing, the turna fell and plunged into the earth. The baby fell and was covered by the turna, which forced the rocks upward, forming the circular mountain range. The baby's mother, the Evening Star, and father, the Morning Star, continue to search for their baby to this day"* (Image credit: ©Google earth 2012)

Fig. 5.7 Gosses Bluff impact crater in central Australia (Northern Territory) at *23°49'S* and *132°18'E* with a diameter of 22 km and a depth of 180 m was formed in less resistant rock which sustained rather strong weathering, relative to its young age in the geologic time scale (143 million years old). The meteorite may have had a diameter of about 1 km and originally caused a crater depth of 1.5 km. Shatter cones are conical fractures present at this site with typical markings produced by shock waves. The energy of the impact has been calculated to equal about 1 million Hiroshima bombs (Image credit: ©Google earth 2012)

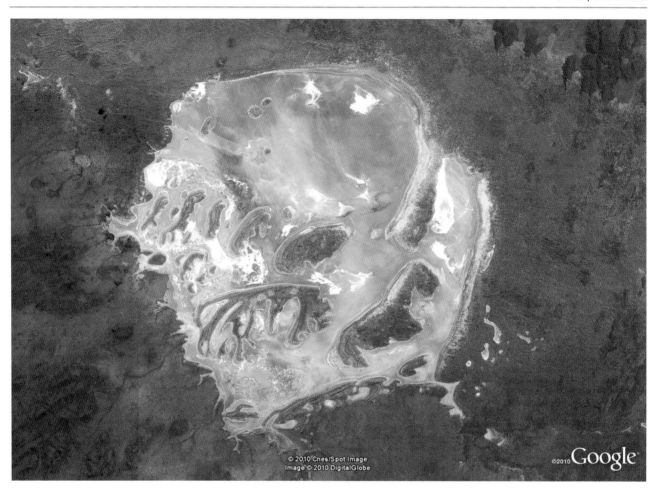

Fig. 5.8 In the arid landscape of Australia, several old impact craters have been detected such as the complex Acraman crater in South Australia with its salt lake at *32°02′S* and *135°26′E*. The crater has a diameter of originally up to 90 km, but today only a ring-like structure close to 35 km in diameter is visible. Its age has been determined to be about 590 million years, but strong erosion has changed all original morphology. Shatter cones in the rock and shocked quartz minerals document the strong impact that has been estimated to equal 5.2 million megatons of TNT (Image credit: ©Google earth 2012)

Fig. 5.9 The Shoemaker crater in Western Australia, at *25°51'S* and *120°54'E* with a diameter of about 30 km, belongs to the oldest impact structures on Earth with an estimated age of 1.63 billion years, although more recent measurements yielded a much younger age of only 568 ± 20 million years. Again shatter cones and the occurrence of shocked quartz document the impact process in Archaean granitic rocks (Image credit: ©Google earth 2012)

Fig. 5.10 The Clearwater impact craters (Quebec, Canada; Clearwater West with a diameter of 36 km, and Clearwater East with a diameter of 26 km) are both located around *56°08′N* and *74°22′W*. It is thought that a binary asteroid around 290±20 million years ago hit the Earth. In the smaller crater (Clearwater east) the central bulge is hidden under water (Image credit: ©Google earth 2012)

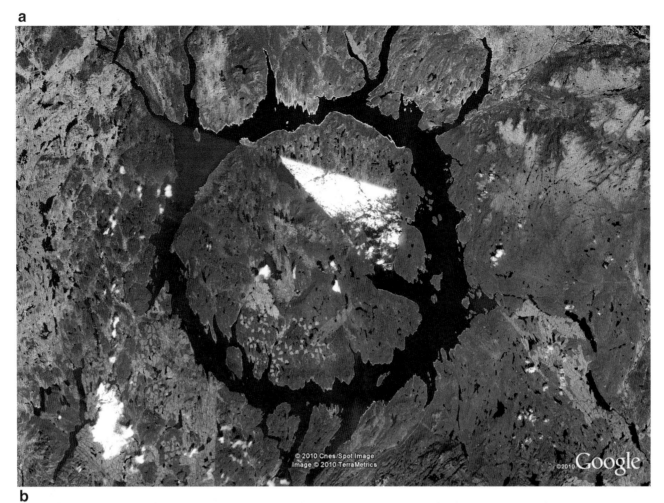

Fig. 5.11 (**a**, **b**) The Manicouagan crater lies at about *51°23′N* and *68°40′W* in Quebec (Canada) with a diameter of 100 km. The crater was most probably formed by an impactor of about 5 km across! The age has been dated to 214 ± 1 million years. The ring-like lake surrounding the central bulge is either an erosional feature (by glacial erosion on less resistant impact breccia), or just lower contour around the central bulge. (Image credit: ©Google earth 2012/D. Kelletat)

Further Readings

Alvarez LW, Alvarez W, Asaro F, Michel HV (1980) Extraterrestrial cause for the cretaceous-tertiary extinction. Science 208:1095

Bobrowsky PT, Rickman H (eds) (2007) Comet/asteroid impacts and human society: an interdisciplinary approach. Springer, New York

Earth Impact Database (2010) Planetary and space science center. University of New Brunswick, Canada. http://www.passc.net/EarthImpactDatabase/

Hamacher DW (2011) Meteoritics and cosmology among the Aboriginal cultures of central Australia. J Cosmol 13:3743–3753

Hamacher DW (2012) On the astronomical knowledge and traditions of aboriginal Australians. Doctor of Philosophy thesis, Department of Indigenous Studies, Macquarie University, Sydney

Koeberl C (2002) Mineralogical and geochemical aspects of impact craters. Mineral Mag 66(5):745–768

Lissauer JJ (1999) How common are habitable planets? Nature 402:C11–C14

Melosh HJ (1989) Impact cratering – a geologic process. Oxford University Press, New York

Piccardi L, Masse WB (2007) Myth and geology, Special Publications, 273. Geological Society, London. doi:10.1144/GSL.SP.2007.273.01.26

Renne PR, Deino AL, Hilgen FJ et al (2013) Time scales of critical events around the cretaceous-paleogene boundary. Science 339:684

Physical and Chemical Weathering

6

Abstract

As one geologist exclaimed: *"Savor the irony should the metamorphic muscles that push mountains to the sky be driven by the pitter patter of tiny raindrops."* In the following chapters we look at this pitter patter, a synonym for exogenic processes that shape our landforms, starting with the most important process: weathering. Weathering takes place through a combination of both mechanical and chemical means. We have all experienced the results of weathering in day-to-day life whether you stumble over a tree root that broke the asphalt on your morning run or you hear the sharp sound of a rock splitting apart in the silence of a freezing mountain night. But the fate of Cleopatra's needle, a collective, popular name of three New Kingdom Egyptian obelisks that were transferred from Egypt to the damp cities of New York, London and Paris during the colonial era of the nineteenth century illustrate that weathering is obviously a strong function of climate, the properties of the parent rock, the presence or absence of soil, and time. This chapter is mostly illustrated with photography because the small-scale details of the weathering process are difficult to capture with images from space. Nevertheless, some large-scale weathering forms may well be detected from space as pictures of exfoliation patterns or corestone weathering show.

In the first chapters we have encountered surface structures from tectonic and igneous processes that constitute endogenic geomorphic processes (driven by forces from within the Earth). In contrast, exogenic processes (solar radiation, temperature changes, wind, water, ice) create relief sculptures on the interface between the atmosphere and Earth's surface. Exogenic processes occur in three main stages: the weathering of rocks, the subsequent transportation of the material and finally the deposition in various sediment sinks. Landforms shaped by exogenic processes are the subject of the following chapters, but we start with the controls on weathering and here we have several key factors to consider: the properties of the parent rock, climate and in particular rainfall and temperature, the presence or absence of soil and the length of exposure to the influences of weathering. The following chapter summarizes some of the most important weathering processes. However, weathering processes often act on very small spatial scales and many result in features too small to illustrate with Google Earth. Photo plates thus help to illustrate some of those aspects of weathering that we cannot explore from satellite images (Figs. 6.1 and 6.2).

The Egyptian obelisks (*obeliskos* meaning a prong for roasting in Greek) also coined Cleopatra's needles (Fig. 6.3) shall be used to visualize the concept of weathering. Around 3,000 years ago, during the Pharaonic period, the Egyptians quarried obelisks of grand size and perfection from the granite quarries near Aswan in Upper Egypt. It is one of the world's most prominent ancient quarry landscapes and the inscribed Egyptian hieroglyphs tell the stories of fame, fate and wars of different pharaohs. For the Egyptians, they were the symbol of a petrified sunray and were erected with the function to perforate clouds and disperse negative metaphysical forces.

Centuries later, the French were the first to transport one of these needles, quarried under the reign of Ramesses II. It was shipped to Luxor, where it apparently stood for more than 3,000 years, and then to Paris' Place de la Concorde in 1836. The twin obelisks of London and New York were both

Fig. 6.1 All weathering processes – and in most cases a combination of mechanical and chemical weathering – work most effectively along existing planes of weakness within the rock structure such as bedding planes and/or joints. (**a**) In Bryce Canyon National Park, Utah (USA), relatively soft and well-stratified sandy limestone units of different colors have been characteristically shaped by retrograde erosion. (**b**) Joints have been carved out in granite rock at Yosemite National Park in California, USA. As pre-defined planes of weakness, they represent starting points for weathering processes (Image credit: D. Kelletat). (**c**) Tree roots may grow along and explore fissures and contribute to the destruction of rock units (Image credit: S.M. May). (**d**) A typical aspect of frost weathering that acts along the parallel foliation planes in these metamorphic schist of the Austrian Alps (Image credit: D. Kelletat). (**e**) A granite boulder located close to the first cataract of the Nile River in Upper Egypt. About 3,600 years ago, a pharaoh ordered to document a famine in the country with these inscriptions in the dark desert varnish of the boulder. Later, mechanical tension due to temperature variations caused a central breakage across the inscriptions (Image credit: D. Kelletat). (**f**) Core-breakage caused by frost wedging in the Little Ice Age moraine of a glacier in southern Iceland (Image credit: S.M. May). (**g**) The process of salt weathering may also result in very large cavities such as this example from the east side of Uluru in central Australia. It is called "The Brain" and the left part has a size of nearly 150 m. The aboriginals of the Anangu People have dreamtime stories about almost every feature of the rock but they keep most of them secret and only pass them onto the appropriate knowledge holders. (**h**) Honeycomb weathering in fine-grained sandstones, southern Crete, Greece (Image credit: D. Kelletat)

Fig. 6.2 (**a**) After several million years of chemical weathering, granite may be altered into sand and only some larger corestones are left. The resistance of granite solely depends on the density of joints. The parts with less joints may still preserve corestones (upper part of the image), whereas below this section in an area of denser joint patterns, even the corestones are completely weathered (Harz Mountains, northern Germany) (Image credit: D. Kelletat). (**b**) Large corestones in the Girraween National Park (southeast Queensland, Australia) that belongs to part of the New England Granite Belt (Image credit: S.M. May). (**c**) A granite outcrop at Wilson's Promontory in southern Victoria (Australia) shaped by chemical weathering processes with large corestones preserved at the top. (**d**) The most important precondition for corestone weathering is a well-developed joint pattern. This image from the Joshua Tree National Park, California (USA) illustrates the initial stadium of corestone formation and its dependence on joints. (**e**) At Uluru (Ayers Rock) in central Australia, the "unloading" by weathering and erosion of former sedimentary layers took place millions of years ago, but hitherto the very old arkose sandstone continues to react to the pressure relief. The reduced pressure leads to a slight expansion of the formerly loaded rock, and the result is a sheet-like fracturing of large rock plates, called "exfoliation" (also "desquamation", *squama* is Latin word for "flake"). The image shows a 2–4 m thick remnant of an exfoliation sheet on the northern slope of Uluru. Fragments of the original sheet can be seen as boulders around the base of the slope (Image credit: D. Kelletat). (**f**, **g**) Exfoliation acting on exposed granite in the Yosemite National Park, California (USA). The region was covered under thick ice sheets until ~12,000 years ago (Image credit: D. Kelletat, S.M. May)

erected in Heliopolis (modern Cairo) during the reign of Thutmose III, and later moved to Alexandria on the coast of the Mediterranean Sea by the Romans at 12 BC. Until its relocation in 1880, the New York obelisk seems to have stood upright throughout nearly 2,000 years since the Roman period; however, its London companion fell over at some stage after the thirteenth century, probably as a result of an earthquake in 1301 BC. The current discussion and controversy about the stewardship of the Cleopatra's needles center around their different state of preservation. The Paris needle is fairing the best in its new Old World home, and the London and New York needles are in increasingly severe deterioration states due to a very complex weathering history that is interestingly explained in Per Storemyr's blog on Archaeology and Conversation (http://per-storemyr.net/). Until recently urban air pollution was seen as the major

Fig. 6.3 Cleopatra's needles – Egyptian obelisks transported to Paris (Place de la Concorde), London (City of Westminster, Victoria Embankment) and New York (Central Park) (Image credit: ©Google earth 2012)

Fig. 6.3 (continued)

cause of such accelerated weathering together with the more humid and wet climate compared to their Egyptian desert origin; however, this view is toned down today in favor of a more holistic perspective that includes the diversity of weathering agents that were most likely active throughout the lifetime of these ancient monuments. This involves tracing the causes of obelisk weathering right back to the geological deposit from where most of them originated – to the granite quarries of Aswan and the varying properties of the stone.

Granite is a light-colored igneous rock composed of large mineral grains that fit tightly together. The dominant minerals are quartz and feldspar together with a variety of other minerals, most commonly the black mica biotite and the black amphibole hornblende. Both mica and hornblende have compositions and crystal structures that favor weathering. Different minerals weather at different rates and differences in the rock's mineral composition and structure can make rocks more or less susceptible to fragmentation and cracking. The most stable minerals (and thus with the slowest weathering rates) are iron oxides, aluminum hydroxides, quartz, clay minerals, micas and feldspar. On the other end of the scale rank the minerals halite, calcite, olivine as the least stable minerals with the fastest rate of weathering. In terms of building for longevity the Egyptians therefore made a wise and probably conscious choice that monuments made from granite will resist weathering and preserve much longer than other rock types such as limestone or sandstone. Still, given enough time, even a weathering-resistant rock or monument will decay and its surface and inscriptions will become blurred without applying any conservation techniques.

Practically, all Egyptian obelisks are damaged at their bases with flakes of granite chipping off since soluble salts from groundwater or the surrounding soil have ascended into the rock matrix over the millennia and, after their relocation, caused accelerated salt weathering in the wetter European and North American climates. Soil retains water and salt and it hosts a variety of organisms, bacteria and vegetation that together promote chemical and physical weathering in a positive feedback process. Water, on the other hand, is relatively unavailable in arid regions such as the Egyptian desert climate and therefore limits the rate of chemical weathering. In generic terms, high temperatures and the high rates of rainfall promote chemical weathering while cold and dry climates slow the chemical weathering process. The longer the time of exposure to weathering, the greater is the impact on the rock in terms of chemical alteration, physical fragmentation or dissolution. The ancient Egyptian quarries with their different rock types are good examples with weathering rates as slow as 0.12 cm per 1,000 years in crystalline rocks. However, they also show that weathering is a very complex phenomenon and that, in contrast to our perception, it is not a linear process over time. As a general observation, rocks weather more rapidly in the tropics than in temperate and cold climates mainly because plants and bacteria grow quicker in warm, humid climates, contributing to the acids that promote faster chemical weathering. But let us explore physical weathering first.

6.1 Physical Weathering

Physical weathering describes all the processes that work without changing the internal structure of the rock-constituting minerals. It can breakdown rock into smaller fragments up to the size of silt. Rocks can break for a variety of reasons, including stress along natural zones of weakness in joints or cracks (Figs. 6.1 and 6.2). The upper part of the lithosphere is brittle and the rocks are susceptible to breakage from any bending, stretching or compression that might occur due to tectonic movements. Breakage also results from the release of confining pressure by erosion of overlying rock, for instance leading to exfoliation patterns (Figs. 6.1 and 6.2). The resulting fractures, called joints, are usually evenly spaced and parallel. However, stress on rocks may occur during several phases with different orientations, resulting in the formation of a number of joint sets of different orientation. Jointed rocks often break into rectangular blocks, or sheets. Joints are also caused by shrinkage forces exerted when lava or magma cools. Columnar jointing, where the rock splits into long columns, is a pattern typically associated with volcanic rocks (see Chap. 2). However, processes of both physical and chemical weathering exploit these joints within the rock. Although classified as biological weathering, the mechanical disintegration of rocks, e.g., through biological activity such as root growth, can be seen as part of physical weathering.

6.1.1 Insolation Weathering

An efficient mechanical process that can breakdown rocks is the expansion and contraction of rocks caused by temperature variations (Fig. 6.1e). The greater the temperatures vary and the more frequent temperature changes take place, the more mechanical breakage will occur. The desert climates are a good example where temperatures are high during the day, but may often drop below 0 °C during the night. This temperature-depending process is termed "insolation weathering" (*sol* in Latin means sun).

Insolation weathering is determined to a large degree by the color of the rock surface. Lighter colored rocks such as granite will reflect the incoming radiation more in comparison to darker rocks (e.g., basalt) which absorb more radiation and thus heat up and cool down more significantly between day and night. Breakage will result from the expansion and contraction of the rock and will be most effective in the outer layer of the rock surface. The mechanical tension between the outer, weathered and fresher, unaltered rock matrix can also contribute to fracturing of the rocks. In addition to the normal warming and cooling cycle from day to night, a warm rock surface may be cooled down by a cloud's shadow or by cold shower of rain – thus, the cycles of expansion and contraction may occur several thousand times during 1 year.

6.1.2 Frost Wedging and Salt Weathering

Mechanical breakage of rock is also prominent in colder and mountainous climates where water episodically freezes and thaws. As water freezes, it expands about 11 % in volume and is thereby exerting a force strong enough to open a crack or a fracture in a rock. Frost wedging is active if temperatures fluctuate just around 0 °C. Mechanical disintegration by frost wedging is largely dependent on how frequent water changes from its fluid state into ice – the more often this takes place, the more breakage will occur.

Similar effects can be attributed to salt weathering, where salt crystallize from saline solutions in cracks and joints and the expansion of the salt crystals exerts pressure and contributes to the disintegration of the rock.

6.1.3 Exfoliation

Exfoliation belongs to a type of physical weathering but is not directly related to the interaction between climate and the lithosphere. The term exfoliation is a process in which large flat or curved sheets of rock are detached from, in most cases, crystalline rock bodies. Together with corestone weathering (see below), it is one of the few weathering features that are easily visible using Google Earth (Fig. 6.4). The fractures form when near surface rocks experience high residual compression parallel to the surface, followed by pressure relief, for instance caused by the buildup and melting of glaciers during glacial periods or the erosion of thick overlying sediment layers.

6.2 Chemical Weathering

6.2.1 Hydration, Hydrolysis and Oxidation

Chemical weathering occurs when minerals react with water and air. In the process, some minerals may dissolve whereas others combine with molecules such as oxygen, water or carbon dioxide to form new minerals. The chemical attachment or incorporation of water molecules to or into a particular mineral (hydration) may lead to an increased volume of the minerals. For example, the mineral anhydrite ($CaSO_4$) combines with water (H_2O) to form gypsum ($CaSO_4 \cdot 2H_2O$). Hydration may also be classified as a physical weathering process due to the mechanical pressure caused by the increased volume. Oxidation is also important as a process of weathering, mainly affecting iron and manganese components

Fig. 6.4 The Enchanted Rock granite dome is part of the Town Mountain Granite batholith in Texas (USA) and it shows characteristic exfoliation patterns which can be observed effortlessly with Google Earth (Image credit ©Google earth 2012)

in silicate minerals. Contact with O_2 causes changes in the oxidation state of these cations at the minerals surface, forcing them to be segregated from the crystal lattice. Manganese-bearing iron oxides and hydroxides often occur as goethite, hematite (ferric oxide) and are responsible for the brown and red colors of soils and weathered rocks.

The most important process in chemical weathering is hydrolysis, which involves structural changes or even complete dissolution of the weathered minerals. The dissociation of H_2O forms H^+ and OH^- ions that chemically react and recombine with the crystal lattice of minerals by ion substitution. Hydrolysis is common for silicate minerals and results in the destruction of primary minerals (e.g., feldspar) and the formation of secondary clay minerals (e.g., kaolinite), during which large amounts of silica is washed away, discharged, or dissolved. In tropical climates, hydrolysis is responsible for the development of thick sheets of weathered material, called saprolite.

A typical aspect of chemical weathering in granite landscapes is the development of corestones or "woolsacks": Because granite is so massive, chemical weathering occurs almost exclusively on joint surfaces resulting in a soil profile and a ground surface containing numerous corestones (Fig. 6.2). The presence of joints allows water to reach the otherwise impermeable rock. Thus the more jointed the rock, the higher the rate of chemical weathering as the joints increase the available surface area exposed to water. Due to ongoing weathering along the joints and particularly joint intersections, the unaltered material in the middle of a jointed block takes on a spherical shape referred to as corestones. The weathered material next to the joints is susceptible to further weathering and may even be washed away, leaving the spherical corestones behind (Figs. 6.2, 6.5, and 6.6).

6.2.2 Tafoni or Honeycomb Weathering

There are other, more complex expressions of chemical weathering in jointed rocks such as honeycomb weathering and *tafoni* (plural: *tafone*) (Fig. 6.1). The word comes from Sardinia, where spectacular honeycomb structures form in the coastal granites. About 3,500 years ago, the Minoans first documented the stages of tafoni evolution in magnificent fresco paintings. Tafoni development combines physical weathering (salt dissolution and recrystallisation) with chemical weathering (hydration) and affects a variety of rocks in a range of environments. Tafoni weathering patterns evolve in distinct stages, from tiny pits to large caves. Tafone may have arch-like entrances, smooth, concave walls, and sometimes floors covered with rock debris. They may develop on building stones and they may shape ocean cliffs, rocks in hot deserts or arctic landscapes. Although it was first noted in the

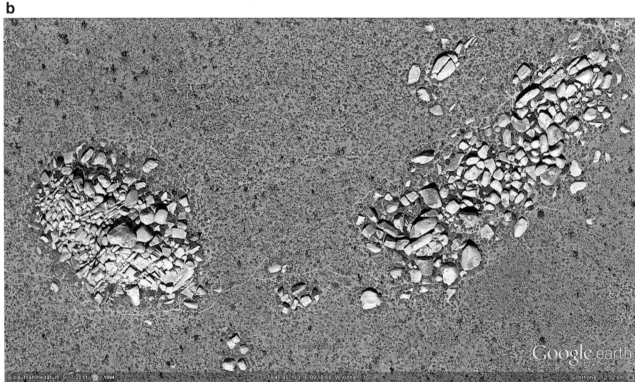

Fig. 6.5 (a) Within thick granite plutons of the Joshua Tree National Park, different joint patterns in the eastern and western section of the image are responsible for the different size of boulders assembling at the surface. The formation of these typical corestone landscapes is caused by "corestone" or "woolsack" weathering. (b) Corestone clusters are left behind as remnants of disintegrating thick granite plutons at the Joshua Tree National Park, California (USA) (Image credit: ©Google earth 2012)

6.2 Chemical Weathering

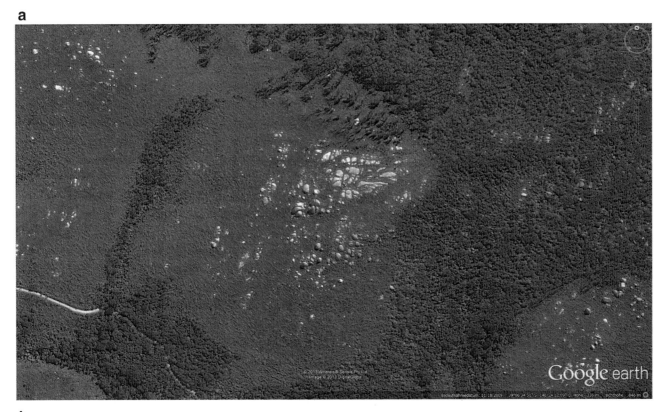

Fig. 6.6 (a, b) In southern Victoria (Wilson's Promontory National Park, Australia) large corestones can be detected from space. In the two images, the subsurface structure of the granite including joint patterns from which the corestones originate is visible. The corestones may reach diameters of 10 m and greater (Image credit: ©Google earth 2012)

nineteenth century, its origins are still poorly understood, and a dearth of laboratory experiments in testing the many proposed mechanisms for its development has added to the ambiguity. Experiments show that heterogeneous wind flow over a rock surface is important in the development of this weathering pattern. Wind promotes evaporative salt growth (and thus mechanical pressure) between grains on a rock surface, resulting in the development of small, randomly distributed cavities. A reduction in air pressure within the cavities results in increased wind speed and rapid evaporation. The comparable high evaporation rate and evaporative cooling of the saline solution in the cavity leads to increased granular disintegration compared to the surrounding areas and result in the formation of honeycomb features. A famous tafoni feature is 'The Brain' on Uluru in central Australia (Fig. 6.1). Often you will find tafone developed at the base of rock formations. This is due to salt weathering acting on the more shaded and humid parts of the rock.

Further Readings

Atkinson D (2004) Weathering, slopes and landforms. Hodder Arnold, London

Bahat D, Grossenbacher K, Karasaki K (1999) Mechanism of exfoliation joint formation in granitic rocks, Yosemite National Park. J Struct Geol 21:85–96

Bland W, Rolls D (1998) Weathering: an introduction to scientific principles. Hodder Arnold, London

Curran BA, Grafton A, Long PO, Weiss B (2009) Obelisk: a history. MIT Press, Cambridge

Doehne E (2002) Salt weathering: a selective review. Geol Soc Am Bull 74:519–527

Goudie AS, Viles H (2008) Weathering processes and forms. In: Burt TP, Chorley RJ, Brunsden D, Cox NJ, Goudie AS (eds) The history of the study of landforms or the development of geomorphology, vol 4. Geological Society, London, pp 129–164

Hall K (1999) The role of thermal stress fatigue in the breakdown of rock in cold regions. Geomorphology 31:47–63

Hall K (2004) Evidence for freeze-thaw events and their implication for rock weathering in northern Canada. Earth Surf Process Landf 29:43–57

Hall K, André MF (2001) New insights into rock weathering as deduced from high-frequency rock temperature data: an Antarctic study. Geomorphology 41:23–35

Harrell JA, Storemyr P (2009) Ancient Egyptian quarries – an illustrated overview. In: Abu-Jaber N, Bloxam EG, Degryse P, Heldal T (eds) QuarryScapes: ancient stone quarry landscapes in the eastern Mediterranean, Special Publication 12. Geological Survey of Norway, Trondheim, pp 7–50

Mandl G (2005) Rock joints. The mechanical genesis. Springer, Berlin

Nash DJ (2001) Arid geomorphology. Prog Phys Geogr 25:409–427

Rodriguez-Navarro C, Doehne E (1999) Salt weathering: influence of evaporation rate, supersaturation and crystallisation pattern. Earth Surf Process Landf 24:191–209

Romani JR, Twidale CR (1999) Sheet fractures, other stress forms and some engineering implications. Geomorphology 31:13–27

Storemyr P (2010) Cleopatra's needle: tracing obelisk weathering with old photos. http://per-storemyr.net. Accessed 25 Mar 2013

7 Karst Landscapes: Topographies Sculptured by Dissolution of Rock

Abstract

Water is essential to a wide variety of geological processes, both as a solvent of minerals in rock and soil and as a transport agent that carries away dissolved and weathered materials. The development of all karst landforms requires the presence of rock types capable of being dissolved by surface water or ground water. Throughout the world, karst landscapes vary from rolling hills dotted with sinkholes to jagged hills, pinnacle karst and tower karst found in the tropics that are often featured in Hollywood films due to their breathtaking scenery (think back to 1974 and James Bond in *The Man with the Golden Gun*). Karsts and their associated cave systems have also provided us with some of the most important discoveries of humanity's past, such as the rock art paintings of Lascaux and Chauvet in France and the remains of our ancestors in limestone caves of South Africa. Understanding cave and karst systems is important because 10 % of the Earth's surface is occupied by karst landscape and as much as a quarter of the world's population depends upon water supplied from karst areas.

The term karst describes a distinctive topography and scenery that is shaped by the dissolution (also called chemical solution) of soluble rocks by surface or ground water. Karst landforms range in scale from microscopic (e.g., chemical precipitates) to entire drainage systems and broad karst plateaus covering hundreds of square kilometers. Over a very long period of time, all rock types may undergo chemical solution, but only in a small group of sedimentary rocks dissolution dominates all other geomorphic processes and results in typical morphologies. These are chemical and biogenic sedimentary rocks that are dominated by the minerals calcite ($CaCO_3$; i.e. forming limestone), dolomite ($MgCaCO_3$), gypsum ($CaSO_4 \cdot 2H_2O$), anhydrite ($CaSO_4$), and halite ($NaCl$). Over the same time scale, the amount of dissolution of these different rocks is equal to 1 part of limestone to 100 parts of gypsum to 10,000 parts of salt. It is evident that in humid climates dissolution occurs faster than in arid regions, where only dew may be available as a solution agent. Thus, typical limestone karst landscapes are most abundant in humid regions. In contrast, karst in sedimentary rocks composed of halite is encountered only under very arid climate conditions and environments, since frequent precipitation would rapidly dissolve the salt without preserving the distinct surface morphology (see also Figs. 4.36 and 4.37a).

Rain or meltwater becomes weakly acidic by reacting with atmospheric carbon dioxide (CO_2) to carbonic acid (H_2CO_3). When this water seeps through rock fractures or soil, it may pick up additional CO_2 from plant roots, microbes, and other soil-dwelling organisms. The acidity in the water tackles the $CaCO_3$ minerals of the bedrock particularly along rock surfaces of joints or fractures, forming calcium bicarbonate or calcium hydrogen carbonate $Ca(HCO_3)_2$. In contrast to calcite, the calcium bicarbonate is water-soluble and is removed with the water. The chemical reaction describing limestone dissolution is:

$$CaCO_3 + CO_2 + H_2O = Ca^{2+} + 2HCO_3^-$$

Eventually a network of chambers, cave passages and caverns will form in the limestone. Thus, the term karst also describes an aquifer type which is characterized by an underground network of conduits and caves formed by chemical dissolution of the surrounding carbonate rocks.

The formation of such networks generally takes place in both the vadose (unsaturated zone, zone free of water) and the phreatic zone of the karst aquifer, where all voids are filled with water under pressure greater than atmospheric. In this zone, a phenomenon called "mixed corrosion" is important for the progressive solution of cave systems – two saturated water flows in the karst aquifer merge and the mixed water is capable of dissolving the limestone again. However, cave formation is most extensive in the epiphreatic zone, where unsaturated water from flood-related fluctuations of the karst's water table is most corrosive. Large drainage systems in karst areas are likely to have both fluvial (surface) and karst (underground) drainage components.

In a warm rainy climate with 1,000 mm of precipitation per year, water could dissolve limestone at a rate of 1.8 cm per 1,000 years. As cold water has a higher potential of becoming more acidic (because cold fluids can hold more CO_2; compare your champagne bottle!), karst is also well developed in cold humid areas at mid-latitudes. For example 10 °C water dissolves 2 times more CO_2 than 30 °C water and 0 °C water dissolves 3 times more CO_2 than 30 °C water.

The local environment on the surface is an important factor for the development of surface karst forms: on bare rock, solution can only act as long as the surface of the rock is wet. The associated landform is called the "naked karst", which shows in particular very sharp small linear rill forms (Fig. 7.1a). Under a soil and vegetation cover, the contact zone to the rock may be constantly wet and dissolution no longer depend only on the frequency of rainfall. We call this "covered karst", and the resultant forms are rather smooth. These karst forms are organized in several categories. The most minor but widespread surface karst landforms are small solution pits, grooves and runnels, termed *karren* with dimensions of only several millimeters up to a few centimeters. Large-sized mega-karren are recognizable within a 10–100 m grid. Mega-karren landscapes may further be transformed into pinnacle karst landscapes such as the one developed in Madagascar (Fig. 7.1).

Fig. 7.1 (a) Typical karren features on a limestone rock in Greece. They are about 50 cm long and distinct in shape at the upper parts. Rainwater is acidic and erosive and it becomes increasingly more saturated with soluble carbonate and consequently less erosive as it flows downwards (Image credit: D. Kelletat). (b) Intersecting karren features tracing different joint patterns in limestone of the Aran Islands, Ireland (Image credit: A. Scheffers). (c) Karren landscape in the Judbarra/Gregory National Park (Northern Territory, Australia), developed in the carbonate and calcareous rocks of the Proterozoic Skull Creek Formation. A clear zonation of karren formation resulting from

Fig. 7.1 (continued) progressively longer surface exposure after slope retreat is visible. The eastern (right) part of the image shows incipient (and covered) karren formation, whereas advanced mega-karren formation has dissected the western part (east of the East Baines River channel) (*16°03′54″S, 130°22′49″E*) (Image credit: ©Google earth 2012). (**d, e**) Mega-karren pattern and pinnacle karst in Madagascar. The karst features formed along tectonically induced cracks and joints (Image credit: ©Google earth 2012)

Fig. 7.1 (continued)

7 Karst Landscapes: Topographies Sculptured by Dissolution of Rock

In some places, roofs of limestone caves can collapse suddenly due to subsurface dissolution and result in a collapse sinkhole – a steep-walled depression on the land surface above a cavernous limestone formation (Figs. 7.2, 7.3, and 7.4). A sinkhole may also develop over time without a sudden collapse where joints intersect each other and dissolution is highest (solution sinkhole). Sinkholes (or *dolines*) are characteristic of a distinctive type of topography known as Mediterranean Karst, which is named for a region in the northern part of former Yugoslavia (Fig. 7.3a, b), but

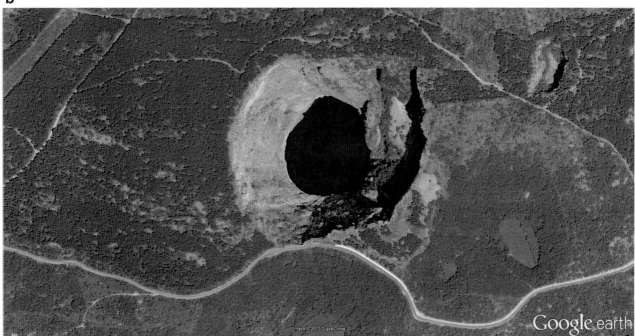

Fig. 7.2 (a) Two large sinkholes with diameters of 85 m in the southern Peloponnese near Didyma (Greece). Note the size of soccer field in the right part of the image as scale. (b) The Crveno Jezero ("Red Lake") close to Imotski (Croatia) is one of the world's largest collapse sinkholes, reaching down to the karst water table. (c) A field of sinkholes west of the Crveno Jezero, some reaching down to the karst water table. The doline field lies at the western margin of a large polje, of which large parts are used for agriculture (lower left part of image). The westernmost part of the flat polje floor is flooded during winter due to rising karst water tables (Image credit: ©Google earth 2012)

c

Fig. 7.2 (continued)

a

Fig. 7.3 (a) Typical karst landscape north of Raško Polje at *43°35′N* and *17°10′E* (Bosnia and Herzegovina) characterized by numerous sinkholes. Width of scene is about 8.5 km. (b) Deeply dissected karst plateau in northern Montenegro showing the typical pattern of Mediterranean Karst with abundant sinkholes (*43°13′N, 18°57′E*). Adjoining sinkholes have formed uvalas particularly in the southeastern part of the image. (c) A karst plateau with thousands of sinkholes in Bosnia and Herzegovina (*43°40′N, 18°05′E*). At several places, particularly in the image center, series of adjoining sinkholes have converged to form uvalas. (d) A sinkhole landscape in Morocco at *28°11′N* and *11°20′W*. Image width is about 12 km (Image credit: ©Google earth 2012)

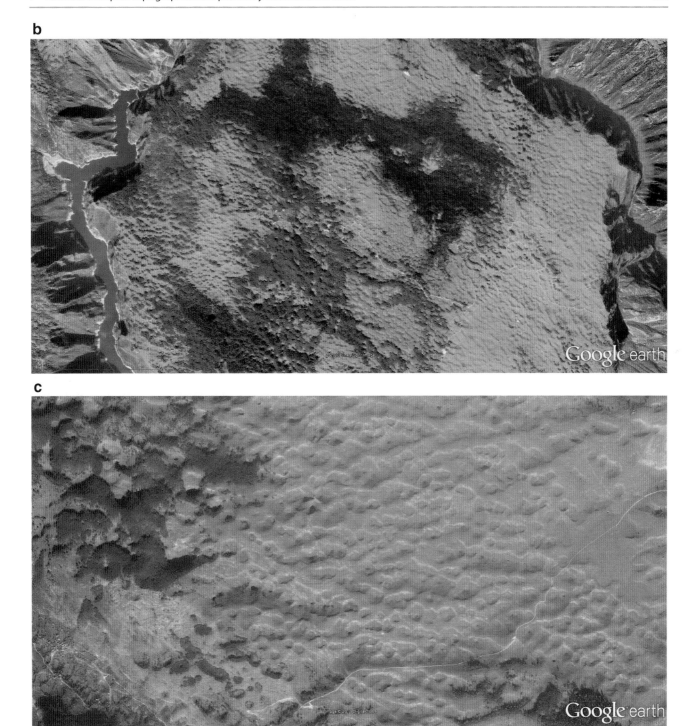

Fig. 7.3 (continued)

d

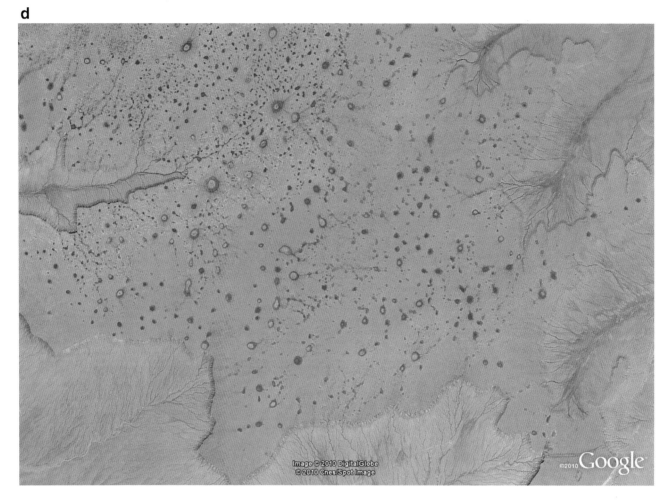

Fig. 7.3 (continued)

sinkholes dominate numerous landscapes worldwide. Underground drainage channels replace the surface drainage system of small and large rivers and short, rare streams often end in sinkholes, detouring into the underground and sometimes reappearing kilometers away.

The form of sinkholes may vary from smooth depressions to vertical shafts, especially if a subterranean cave has collapsed. Blue holes within coral reefs are sinkholes, which developed in limestone when sea level was lower during the last glacial periods. Thousands of sinkholes can be found within the Nullarbor Plain (built of limestone) in South and Western Australia (Fig. 7.4), and in cemented carbonate sands of former dunes along the coastlines of Victoria and southern Western Australia which exhibit beautiful caves systems.

Sinkholes can coalescent along main joints into *uvala* (a Serbian-Croatian word from the classic Yugoslavian karst region), which is often elongate in form (Fig. 7.3c). A series of uvala may also mark a former subterranean stream channel but does not necessarily contain a stream at the present time. An uvala is generally of the order of 1 km in length and thus is intermediate in size between a sinkhole and a *polje*.

Poljes (Serbian word for "field") are found where tectonic depressions have developed in limestone areas and are subsequently enlarged by solution weathering along their margins. They typically do not have superficial drainages but are drained by so called *ponors*, where water is flowing into subterraneous karst systems (Fig. 7.5). Large parts of the sedimentary infill that creates the flat surface topography of poljes is derived from insoluble material in the limestone, with clay or silt contents. Together with the lack of surface drainage systems, this relatively impermeable sedimentary cover leads to extensive flooding in many poljes during winter, when the karst water table rises (see also Fig. 7.2c). Throughout the dry season, poljes are places of extensive farming, while large-scale flooding of the same polje allows for fish cultivation during winter. Moreover, the sediment cover essentially protects the geologic basement of the poljes from further dissolution and promotes lateral extension.

Among the most distinctive and unique landscapes in the world are the tropical cockpit karst landscapes with their scenic topography of rounded peaks and steep-sided, star-shaped depressions. Jamaica is the type area for cockpit

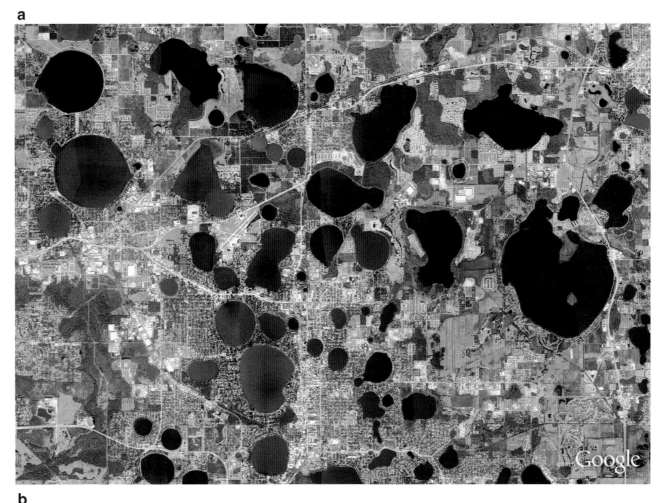

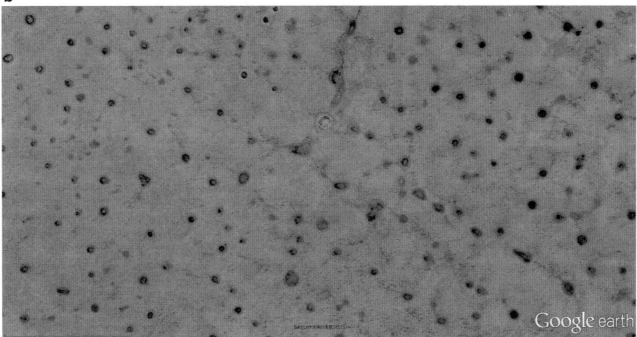

Fig. 7.4 (a) Large sinkholes filled with groundwater in the carbonate rocks of the Florida peninsula (USA) at about *28°03′N* and *81°41′W*. Image width is 20 km. (b) Numerous sinkholes at *30°18′S, 129°10′E* (Nullarbor Plain, South Australia). (c) The patchy karst landscape at Nullarbor Plain (South Australia) (Image credit: ©Google earth 2012)

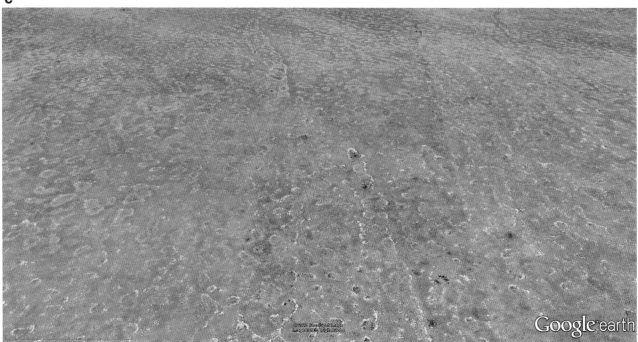

Fig. 7.4 (continued)

Fig. 7.5 (**a**) The polje of Avezzano in the central Apennines of Italy is extensively used for agriculture (diameter of about 17 km at *41°59′N* and *13°33′E*). (**b**) Small polje in southern Slovenia (*45°35′N, 15°02′E*). In the foreground, the landscape is characterized by sinkholes. (**c**) The

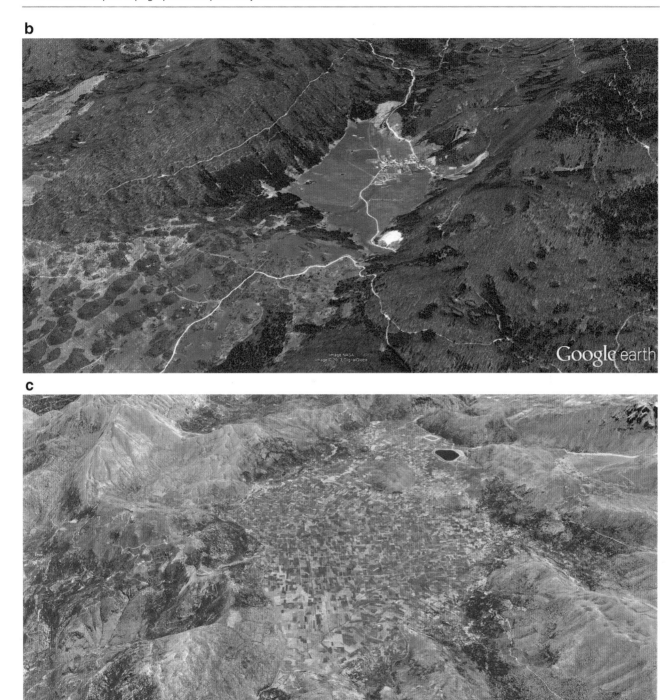

Fig. 7.5 (continued) extensively farmed area of the Lassithi Polje in Crete (*35°10′N, 25°27′E*). The surficial drainage channel disappears into the subterranean karst water system through the ponor in the north-western part of the polje (lower left center of the image) (Image credit: ©Google earth 2012)

karst landscapes (Fig. 7.6). Sufficient widening of these depressions may create a lower-level plain from which the remnants of the limestone blocks stand out as isolated, steep or near-vertical towers, termed cone and tower karst. The cones along the Li River in Guilin (China) tower hundreds of meters above the surrounding plain. The scenic tower karst landscapes in southern Thailand and the South China Sea are beautiful examples of this karst type (Figs. 7.7 and 7.8).

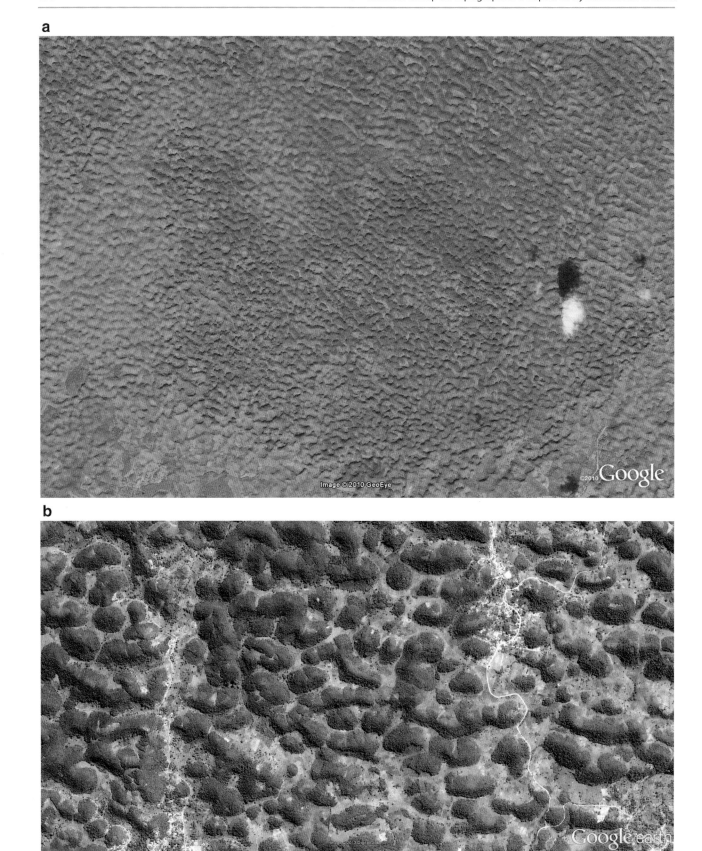

Fig. 7.6 (a) Cockpit karst landscape as a typical feature of tropical karst in the Dominican Republic on Hispaniola Island (Caribbean; *18°57′N and 69°37′W*). (b) Detail of cockpit karst in Jamaica at *18°18′N, 77°17′W* (Image credit: ©Google earth 2012)

7 Karst Landscapes: Topographies Sculptured by Dissolution of Rock

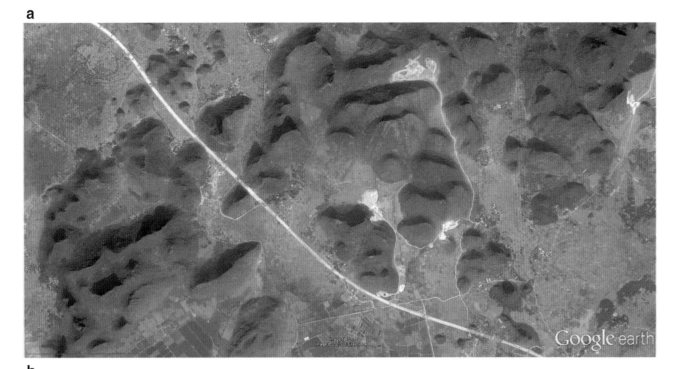

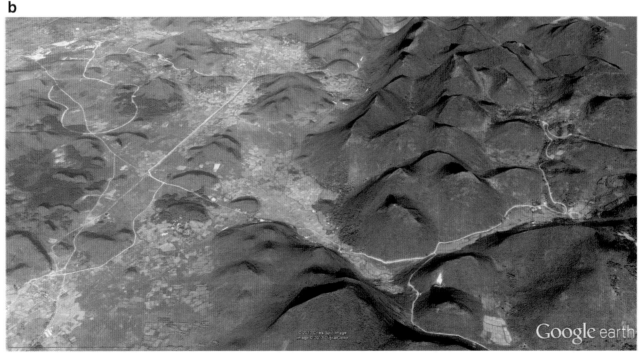

Fig. 7.7 (a, b) Different aspects of the famous cone karst in the area of Guilin (Guangxi, China). The scenic city of Guilin with its rich cultural history lies amidst limestone towers and cones of different height and shape, ranging from 30 to 80 m tall on the alluvial plain of the Li River. The region experiences a subtropical monsoon humid climate characterized by a sharp contrast between the dry and rainy seasons. Annual mean temperatures of 18.3 °C and an annual precipitation of 1,936 mm coincide with the thresholds of temperature and precipitation for subtropical karst. The Chinese people call this landscape *'fenglin'* which translates to 'peak forest' – a more poetic equivalent of the term cone karst (Image credit: ©Google earth 2012)

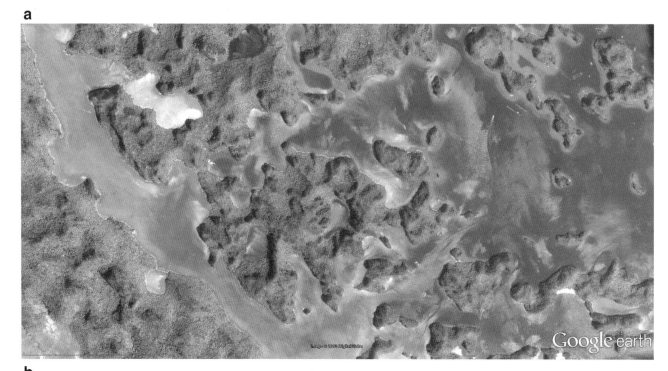

Fig. 7.8 (a) Tower karst in the South China Sea at about *20°46′N* and *107°03′E*; image width about 40 km wide (Image credit: ©Google earth 2012). (b) Tower karst in Phang Nga Bay (southern Thailand, *~8°13′N* and *98°30′E*). The Cuban term "*mogotes*" describes similar residual karst mountains in the tropical karst landscapes of the Caribbean. Some areas exhibit extremely regular conical hills and cockpits and it is hard to believe that these landscapes have been formed just by chance by natural processes alone (Image credit: D. Kelletat)

Well-known features of karst environments are caves and underground rivers. Rivers disappear through joints into the limestone groundwater and slowly dissolve its way into and intricate systems of cave chambers. Throughout the history of humankind, caves have been used as dwellings, refuges, and places of worship for humans and these caves are shared amongst other highly adapted cave-dwelling species. In a way, caves are nature's time capsules as cave sediments, both clastic deposits of clays, sands and gravels and speleothems (chemical cave deposits – mostly calcium carbonate) preserve evidence of past life and climates. This opportunity is only available in a few surface habitats, where weathering processes and erosion are constantly removing the record of past environments.

Where slowly dripping saturated water exits limestone joints or cracks and contacts with the air inside the cave, some of the $Ca(HCO_3)_2$ in solution will precipitate back into $CaCO_3$. Stalactites on the cave ceiling are formed slowly by continuous $CaCO_3$ precipitation. This has nothing to do with evaporation as the relative humidity in caves is often close to 100 % – it is simply the oversaturation of the water, forced by sudden temperature increase and/or pressure decrease. It can take a very long time for most stalactites to form; they usually "grow" less than a millimeter a year and only some centimeters every century. The water dripping from the end of a stalactite falls to the floor of a cave and precipitates more $CaCO_3$ into a mound forming a stalagmite over time. The aesthetic tapestry of these delicate speleothems can be used as geologic archives, capturing thousands of years of past climate history. The Carlsbad Caverns in New Mexico (USA) and the Mammoth Cave in Kentucky (USA) are famous for their architecture of stalactites and stalagmites.

Further Readings

Ford DC, Williams PW (2007) Karst hydrogeology and geomorphology. Wiley, Chichester

Frumkin A (ed) (2013) Karst geomorphology. In: Shroder J, Frumkin A (eds) Treatise on geomorphology, vol 6. Academic Press, San Diego

Grimes KG (2012) Surface karst features of the Judbarra/Gregory National Park, Northern Territory, Australia. Helictite 41:15–36

Márton V, Lóczy D, Zentai Z, Tóth G, Schläffer R (2008) The origin of the Bemaraha tsingy (Madagascar). Int J Speleol 37:131–142

McDermott F (2004) Palaeo-climate reconstruction from stable isotope variations in speleothems: a review. Quat Sci Rev 23:901–918

Mass Movements: Landforms Shaped Under the Force of Gravity

8

Abstract

"A heavyweight for Einstein – Probing gravity where no one has done it before". This was the headline of a press release in April 2013 by the Max Planck Institute for Radio Astronomy (Bonn, Germany). John Antoniadis, a student of the institution and his colleagues put Einstein's general theory of relativity to the test in a cosmic laboratory 7,000 light years from Earth, where two exotic stars – a pulsar or neutron star and a white-dwarf – are circling each other. Einstein's theory states that objects with mass cause a curvature in space-time, which we perceive as gravity. And the scientists' results matched Einstein's prediction perfectly even in the most extreme conditions tested yet. It is almost impossible to imagine that the gravity on the pulsar's surface is 300 billion times greater than the gravity on Earth. From all four universal forces of nature, gravity is the weakest one and equates 9.81 m/s² close to the surface. You can overcome it easily for example if you drop an iron nail on the floor and pick it up with a small bar magnet from a child's toy. But gaze out of the window: Gravity is shaping the landscape everywhere around us when it exceeds the cohesive strength of slope materials and friction and as a consequence downslope movements of masses of soil, sediments or rocks occur. Collectively they are termed mass movements. The downslope movement can be very slow and almost imperceptible or very sudden and catastrophic when massive landslides leave a path of destruction in their wake. This chapter shows examples of different mass wasting processes and landforms.

Gravity acts everywhere in the universe and everywhere at the Earth's surface, pulling everything towards the center of the Earth. But what really causes the force of gravity to take over and initiate slope materials to either fall, slide or flow, and thereby shaping the morphologies of mountain and valley systems, both on the continents and on the ocean floors? To answer that question we have to consider the three most important parameters that influence mass wasting by lowering the resistance to movement.

1. The *steepness* of the slope reflects its state of weathering and fragmentation of the bedrock as well as the geologic-tectonic preconditions. Slopes built up from hard rocks like massive granite or limestone that form sheer vertical cliffs. In contrast, slopes in softer rocks such as shales or volcanic ashes are more prone to weathering and are easily fragmented – they become unstable and slide downhill thereby assume to an equilibrium slope called the angle of repose. For unconsolidated materials, this angle of repose increases with grain size up to about 37°. At any slope that is steeper than the angle of repose and thus unstable, gravity will take over and lead to mass wasting processes that reinstall the stable angle of repose for the specific material.

2. *Consolidated vs. unconsolidated* slope material. Slope materials vary greatly depending on the geology of the local terrain. The stability of a slope depends largely on the rock type or constitution of the material that builds the local terrain. Slopes can be built up from unconsolidated materials (clay, silt, sand and gravel), which are loose and not cemented, or from consolidated materials that may be hard bedrock, or materials that are compacted or cemented by binding mineral cements.

3. The amount of *water saturation* in the slope materials. This parameter is defined by the porosity of the slope

material and the amount of pore water, determined by precipitation intensities or other water sources the slope may be exposed to. Both natural and anthropogenic factors may play a role, and an example of the latter is represented by the loss of vegetation cover by deforestation. Water saturation of ground surfaces acts as a lubricant. As a result, the friction between particles or coarser fragments is lowered and movement is more readily initiated – the slope material may start to behave like a fluid and flows under the force of gravity downhill.

As previously mentioned, the precursors to all these mass movement events are weathering processes that weaken the massive bedrock structure along joints, faults, and predisposed bedding planes in sedimentary rocks. Even the pressure of freezing water and in particular repeated freezing and thawing cycles in higher mountain areas, a process called frost wedging, can trigger rockfalls.

Once a slope is unstable, a mass movement is highly likely and all it needs is a trigger to set things in motion. Often mass movements are initiated by ground vibrations such as those that occur during an earthquake or simply by heavy rainfalls during storms or monsoonal weather patterns. Ongoing weathering and erosion processes can gradually steepen slopes and initiate a sudden collapse of the slope flank as well.

Earth scientists classify different types of mass movements according to the nature of the material (coherent rock or unconsolidated material), the velocity of the movement (gradual with a few cm/year or sudden with up to several hundred km/h) and the physical nature of the movement – will the material fall, slide or flow? The latter implies that the moving material behaves as a fluid while sliding masses move more or less as one unit.

However, the initial mass movement type may change into another during the downslope movement and the characteristics of these different phases can vary significantly during one event (e.g., in terms of velocity, depending on the slope gradient and internal composition). Certain mass movements are very dynamic processes and tend to be highly complex, and changes in their physical behavior may occur within one single event.

8.1 Mass Movement of Hard Rock

Let's turn to mass movements of hard rock and start to explore the most basic type: rockfalls (Figs. 8.1 and 8.2). During a rockfall, fragmented and detached materials from steep or vertical mountain slopes plummet to the ground and tumble over the slope in free fall motion with high velocity,

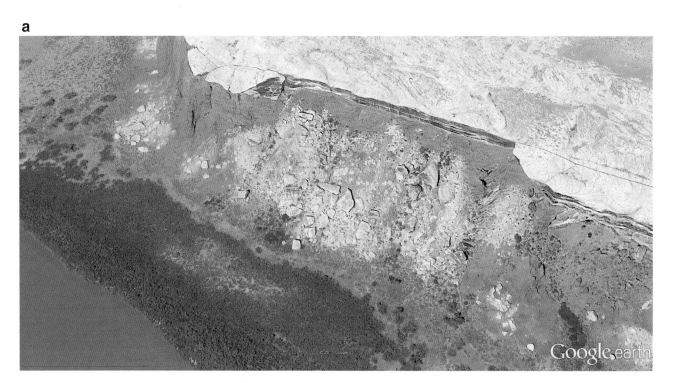

Fig. 8.1 (a) Rockfall of single large boulders along the "White Rim" of Canyonlands National Park in Utah, USA. Large parts of the softer underlying rock layers are covered by block- and boulder-sized rock debris. Their shape and size is predefined by the joint pattern in the overlying carbonate sandstone. Image shows a similar location as in Fig. 8.14b. (b) A larger section of the White Rim in Canyonlands National Park in Utah, USA. Numerous blocks below the White Rim escarpment demonstrate the importance of rockfalls for the back-wearing of the carbonate sandstone and, ultimately, the canyon. (c) Rockfall in Yosemite National Park, California, USA. The mass movement may

b

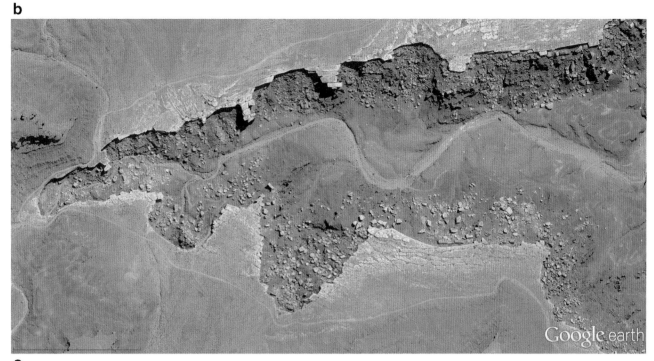

c

Fig. 8.1 (continued) also be termed as a rock avalanche, since rock volume is relatively high and the rock debris has reached the opposite valley slopes, pointing to a more complex process of downslope movement *38°05′20″N, 119°26′25″E*. Traces of an older rockfall or avalanche can be seen in the right (northern) part of the image, where scarce vegetation marks the pathway of the rock debris (Image credit: ©Google earth 2012)

Fig. 8.2 (**a**, **b**) A rockfall at the western flank of Half Dome, a famous mountain in the Yosemite National Park. The sequence shows the Yosemite Valley before (**a**: January 2005) and after (**b**: July 2012) the event. Large scale weathering (exfoliation) of the granite basement favors the occurrence of rockfalls in the area (Image credit: ©Google earth 2012)

usually over a rather short distance. Ceasing of the movement depends mostly on friction, i.e. the relation of the particle in motion and the roughness of the slope. Over longer time periods the material accumulates at the foot of the mountain in the form of a talus cone (Fig. 8.3a–c). Their profile often reveals a straight slope segment at less than the angle of repose and a basal concavity towards the valley floor. Talus cones formed by rockfalls exhibit a distinct sorting arrangement of their sediments over a longitudinal transect (Fig. 8.3b). Fine particles generally accumulate at the top, coarser particles at the bottom part. Talus cones are important components of alpine sediment cascades as they

Fig. 8.3 Talus cones form below rockfall scars in the upper rock walls and accumulate against the valley bottom. Adjacent talus cones may coalesce laterally if enough rockfall material is delivered to the foot of the mountain slope. However, many talus cones consist of solid rock in their inner part, and only the surface layers are built by rock debris. In the initial stages of talus accumulation, the lower part of the slope/rock wall is protected from weathering and more importantly, from disintegration. In contrast, the unprotected upper slopes are subject to weathering and erosion for a much longer time. Over geologic time the talus neck grows vertically, thereby protecting higher slope areas from collapse. (**a**) Talus cones below almost vertical rock walls in the Val Mora (Graubünden, Switzerland, at *46°34′16.12″N, 10°19′36.11″E*). When rockfall material in the upper cone parts becomes water-saturated, it is transported further downslope by debris flows, forming elongate scars

c

Fig. 8.3 (continued) and lobate deposits in the upper and lower parts of the cones, respectively. (**b**) Talus cones in the Southern Alps of New Zealand (upper Rakaia Valley, at *43°16′1.22″S, 170°58′41.61″E*). The typical sorting pattern with larger blocks in the lower, distal part of the talus cone is visible. (**c**) Talus cones (*right*) and lateral moraines (*left*) south of Bow Lake, Banff National Park, Canadian Rocky Mountains (*51°39′55.29″N, 116°27′29.96″W*). In addition to the rockfall material, debris flows may develop from the talus cones due to water saturation of rock debris sections (Image credit: ©Google earth 2012)

act as both sediment sinks for rockfall debris and sources for debris flows and, ultimately, the fluvial systems.

During *rockslides*, the rock masses slide down a slope and, in some cases, move rather slow (Figs. 8.4, 8.5, 8.6, and 8.7). Large rock units may move as a single unit during rockslides. Internal structural weakness of the bedrock (e.g. along slope parallel bedding planes in sedimentary rocks) is the main reason for rockslides to occur in many cases.

The two major slide types are rotational slides (also called *slumps*) and translational slides (these types also occur in slides of rock debris, see below). Both types may behave very differently, including long-term creep, catastrophic movement that is preceded by long-term creep and sudden catastrophic movement with no creep phase. Rotational slides or slumps are slides in which the downward and outward movement of a mass takes place on top of a concave upward failure surface; they usually move slower than translational slides. A translational slide is a mass that slides downward and outward on top of an inclined planar surface. In particular on steep slopes they can be very fast and destructive, and a major hazard for humans. Rockslides of higher velocity may disaggregate and transform into rock avalanches or debris flows during the process of movement.

Two well-known rockslide examples are the Tschirgant and Köfels events, located in the European Alps. The Köfels rockslide in the valley of the Ötztaler Ache River (Ötztal, Austria) is the largest slide in crystalline rocks in the European Alps (Fig. 8.5a). It is assumed that the friction during the slide generated high temperatures (~1,700 °C!), leading to petrographic changes of the underlying rocks. The slide deposit (dark vegetated area in the valley, Fig. 8.5a) covers an area of at least 11.5 km². It descended from the western slope of the valley (left) damming in transverse the Ötztal Ache River and collided with the opposite side of the valley. The rockslide deposit blocked the mouth of the tributary valley from the east (right), which found a new way of descending as a waterfall into the main valley. Due to the blockage, a 7 km long lake formed, which accumulated 92 m thick of sediments. Simultaneously the river has cut a deep gorge into the slide deposits.

The typical morphological pattern of large rockslide deposits is called Toma landscape (Figs. 8.5 and 8.6) and it is the result of an abrupt stop of the high velocity mass movement. It is characterized by steep and conical Toma-hills (named after a landscape in Switzerland) and is associated to funnel- or basin-shaped depressions, which may have

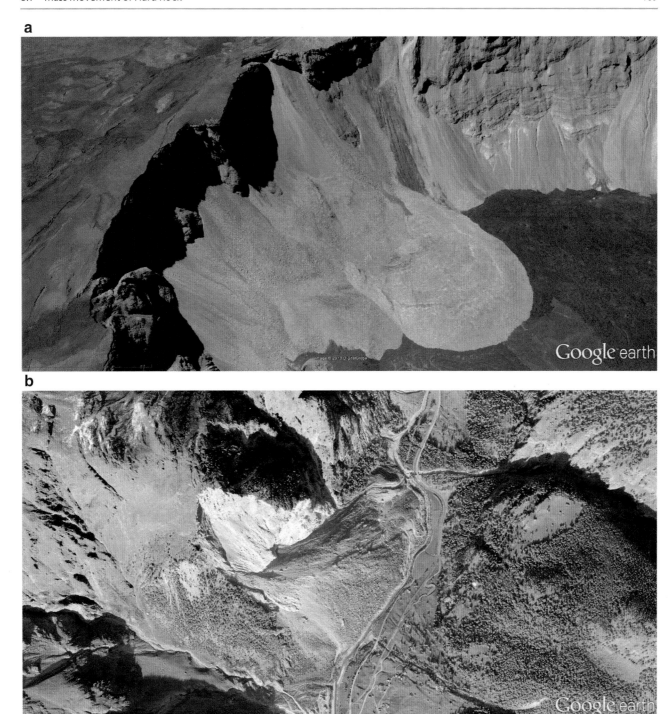

Fig. 8.4 (a) Rockfalls and slides along the walls of a volcanic crater in the Galápagos. The material derived from the crater rim lies on top of relatively fresh and black lava. In contrast to several steep talus cones at the foot of the crater walls, the bright lobate deposit formed during one or several slides. The scar of the larger rockslide is clearly visible on the image. (b) Randa rockslide in Mattertal, Switzerland, with 30 million m^3 of rock in 1991 (Image credit: ©Google earth 2012)

Fig. 8.5 (a) The dark vegetated area in the Ötztal valley (Austria) marks the Köfels rockslide, the largest slide in crystalline rocks in the European Alps. The deposit covers an area of at least 11.5 km². See text for more explanation. (b) Typical Toma landscape at Fern Pass, showing steep and conical Toma-hills, funnel- or basin-shaped depressions transformed into shallow lakes. The Toma landscape (lower image center) was formed after large rockslides ~4,100 years ago; the deep rockslide scar of the northern slide is visible in the upper left part of the image (Image credit: ©Google earth 2012)

Fig. 8.6 The mass movement events (rockslides, sometimes also referred to as rock avalanches) at Tschirgant Mountain occurred along the Inntal fault system, related to internal rock deformation and fracturing. The Toma landscape of the debris is clearly visible (image center). While new chronological information suggests that the events occurred during the 4th millennium before present, interaction of the rockslide with the late-glacial Ötztal glacier was also proposed (Image credit: ©Google earth 2012)

transformed into shallow lake systems between these steep conical hills. A good example is found at the Fern Pass (Austria; Fig. 8.5b), where large rockslides ~4,100 years ago formed the typical Toma landscape.

The Tschirgant Mountain (2,370 m above sea level) is located in the Inn valley of Austria (Fig. 8.6). The massif is composed mostly of dolomites of the Wetterstein Formation and lies at the southern margin of the Northern Limestone Alps. It is separated from the metamorphic Ötztal basement complex by the northeast–southwest-striking Inntal fault system, which causes internal rock deformation and fracturing leading to the event. The Tschirgant rockslide detached from the mountain flank at a height more than 1,400 m (Fig. 8.6, the white rockslide scarp, called "Weißwand", is still visible in the background). It had an estimated volume of ~230 million m^3, spread over an area of 9 km^2. The maximum runout of the rockslide is 6.2 km into the mouth of the Ötztal Ache River, crossing the tectonic and petrographic boundary between the Northern Limestone Alps and the Crystalline Central Alps. The provenance of the rock fragments within the Toma-hills (i.e., the irregular and hilly landscape of the rockslide mass in the valley) gives testimony to this event as they are composed of limestone and dolomite as well as crystalline rocks. During the slides, the rivers Inn and Ötztal Ache were dammed for a considerable period of time as supported by lake sediments in the higher parts of the valleys. Radiocarbon ages of buried wood fragments at Tschirgant suggest two distinct mass wasting events in the 4th millennium before present, though interactions of the rockslide with the late-glacial Ötztal glacier (i.e., the rock mass accumulated on top of the glacier tongue and was (partly) transported down-valley with the glacier ice) was previously proposed.

In addition to large and rapid-moving rockslides with long runout, *rock avalanches* are among the most destructive mass movements of rocks (Figs. 8.8 and 8.9) – they can develop from the fall or slide of a large rock body and involve increasing fragmentation of the rock mass that generates rock fragments of different grain size, further decreasing its cohesion. Rock avalanches travel with extreme velocity (up to more than 150 km/h) over long distances down-valley and can transport large volumes of material. The long travel distances of large flow-like rock avalanches may be explained by the fluidization of the debris caused by the incorporation of air, the existence of trapped air below the moving rock mass, water saturation, and/or additional processes of rock fragment interactions such as dynamic rock fragmentation and granular agitation.

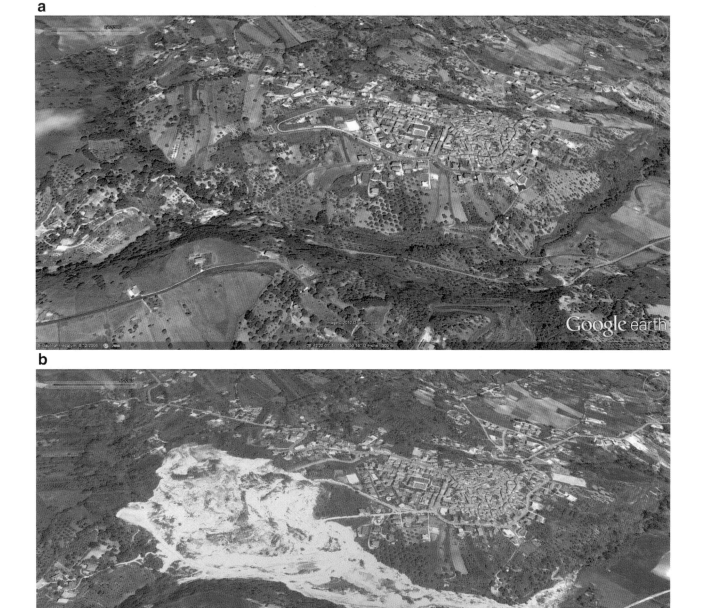

Fig. 8.7 (**a, b**) Sequence of images before and after the Maierato rockslide on February 15th, 2010, in Calabria, southern Italy. The slide was induced by heavy rainfall during the days before the slide, which affected the contact layer between the silt- to clay-rich and evaporitic late Tertiary sedimentary rocks (Image credit: ©Google earth 2012)

Fig. 8.8 Scar of the Val Pola rock avalanche (Valtellina, Italy), which occurred on July 28th, 1987. It destroyed the villages of S. Antonio, Morignone and Piazza. The heavy rainfall and related lateral erosion of debris deposits by the Val Pola River caused a massive slope failure and the motion of more than 35 million m³ rock debris material. Typically, the event underwent several stages: the initial debris avalanche was followed by debris slides, debris flows, and a mud flow, the latter caused by the outflow of a temporary lake, which had formed by damming of the river (Image credit: ©Google earth 2012)

Mass movements such as rock avalanches are typically triggered by earthquakes, but particularly occurred during the transition from the last glacial period to the Holocene, when retreating valley glaciers and ice caps left behind deep valleys with steep slopes. The exposed slopes became unstable under the new ice-free conditions and tended to restore equilibrium conditions of the stable angle of repose. In addition, receding permafrost conditions in the Holocene further destabilized steeper slopes that were bound together by the subsurface ice. Some of the largest alpine rock mass movements have occurred in this period: the Flims rockslide (9–12 km³ rock volume) in the upper Rhine Valley in Switzerland was dated to ~9,500 years before present, and the largest rockslide in the crystalline Alps at Köfels (Ötztal Valley, Austria; Fig. 8.5a) with an exceptional rock volume of more than 2 km³ occurred during the early Holocene, based on radiocarbon dating of buried wood fragments and cosmogenic nuclide dating of surface boulders.

Decaying permafrost during the recent past with warmer than average temperatures is also thought to be a contributing factor to much more recent rockslides such as the Randa rockslides. These rockslides occurred only 3 weeks apart in the Mattertal of the Swiss Alps, involving >30 million m³ of bedrock (Fig. 8.4b).

8.2 Mass Movement of Unconsolidated Materials

Mass movements of unconsolidated materials are slower than most rock mass movements, mainly due to the lower slope angles at which materials like sand, clay, silt or fragmented bedrock or various mixtures of these components become unstable. The moving masses may also include vegetation like trees or man-made materials from infrastructure such as parts of houses, vehicles or fences.

As is the case for hard rock movements, the velocity and the nature of the movement (falling, sliding or flowing) also determines the type of unconsolidated mass movements: The slowest type of unconsolidated mass movements is *debris creep* or *soil creep* which describes a generally gradual and slow downhill movement with a velocity of 1–20 cm per year depending on the slope angle, water content and vegetation density. Vertical tree growth is often offset by soil creep and is visible as typical "hook" growth features at the bottom of the trunk.

Particularly in colder climates on permafrost or in areas with strong winter frost, a certain type of soil creep called *solifluction* occurs, causing typical soil lobes and sheets on gentle slopes (for images see Chap. 13). The water in the top

Fig. 8.9 (a) Rock avalanche deposit (light brown, lower right of image) in Alaska (USA) accumulated on top of the Black Rapids glacier (*63°27′35.33″N, 146°10′0.76″W*). The material derived from the steep wall of the mountain located on the right of the image. The rock avalanche was caused by an earthquake on November 3rd, 2002, occurring along the Denali Fault in Alaska. (b) Another rock avalanche deposit on top of a glacier in the Denali National Park, Alaska (*62°59′6.68″N, 150°30′2.34″W*). The deposit seems to be built up by two separate rock avalanche debris lobes, most likely caused by two different events. The image shows blocks of more than 20 m in length (Image credit: ©Google earth 2012)

soil layer alternately freezes and thaws causing the soil to gradually creep downhill. In permafrost areas, these freeze-thaw cycles are restricted to the "active layer" above the permafrost ground and the movement is restricted to the summer.

Fluid mass movements like *earth flows* composed of relatively fine-grained material such as soil, weathered shales or clay stones, or coarser *debris flows* (Figs. 8.10 and 8.11) occur when high rainfall saturates permeable materials in surface layers that may overly less permeable rocks. If saturated by water, the friction between particles or coarser fragments is lowered, and sections of debris cones may become unstable and start to flow downhill. Where slopes are steep enough this results in a debris flow. Debris flows involve large volumes of sediment, but may flow very fluidly and reach high velocities. The typical central ravines with accompanying coarser ridges are formed due to internal differences in the flowing rock mass, the related pore volume and water contents, and, consequently, the friction within the rock debris, existing between the central and outer parts of the debris flow. While the debris material is still flowing in the center, the rock debris is stopped along the outer parts.

A flowing mass composed of predominantly very fine-grained material (finer than sand) and high water content is termed a *mudflow*. Mudflows and debris flows or avalanches are imminent natural hazards in volcanic regions where rainfall can soak layers of unconsolidated pyroclastic ashes or erupting lava melts huge amounts of snow and ice. In connection with volcanic debris and eruptions, these destructive mass wasting events are called *lahars* (Javanese for lava) (Fig. 8.12).

Similar to rockslides, *debris slides* may travel downhill at lower speeds above one or more failure surfaces (Figs. 8.13 and 8.14). These planar detachments are often related to the existence of colluvial material or rock debris above more competent rocks.

Related to rock avalanches, *debris avalanches* are among the most dangerous and devastating mass movements. Here, in contrast to debris flows, the process of debris movement may pass into sliding or falling. They are common in volcanic and humid mountainous regions and can be characterized, similar to debris flows, by a fluidal behavior. Debris avalanches are very destructive as they can reach extreme velocities. Similar to rock avalanches, the process of movement may be explained by the influence of air, water (e.g., where glacier ice is incorporated into the flow) and rock fragment interactions. In 1962 and 1970, catastrophic ice-debris avalanches occurred on Mt. Huascarán in the Cordillera Blanca of the Andes. The 1970 event was induced by an earthquake and travelled down for 17 km at a velocity of 280 km/h burying the villages of Yungay and Ranrahirca together with more than 20,000 people in Peru.

Fig. 8.10 (a, b) If saturated by water, the friction between particles or coarser fragments is lowered. Sections of debris cones or debris accumulations on the higher slopes may become unstable and start to flow downhill under the force of gravity, resulting in a debris flow. Though containing larger volumes of sediment, debris flows may flow very fluidly. The images show typical associated ravines with accompanying ridges on alluvial fans in northwest Argentina (*23°43′6.40″S, 65°27′2.10″W*) where debris flows are characteristic (Image credit: ©Google earth 2012)

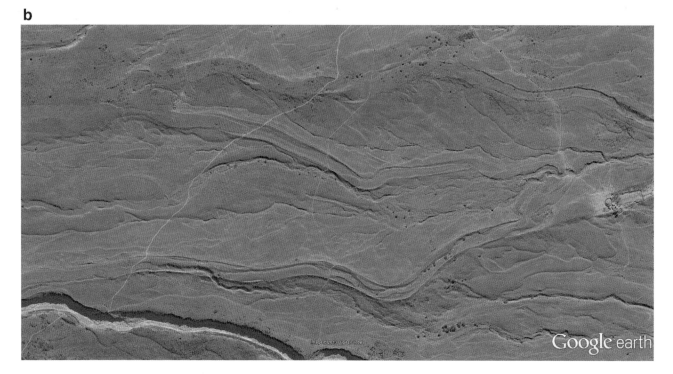

Fig. 8.10 (continued)

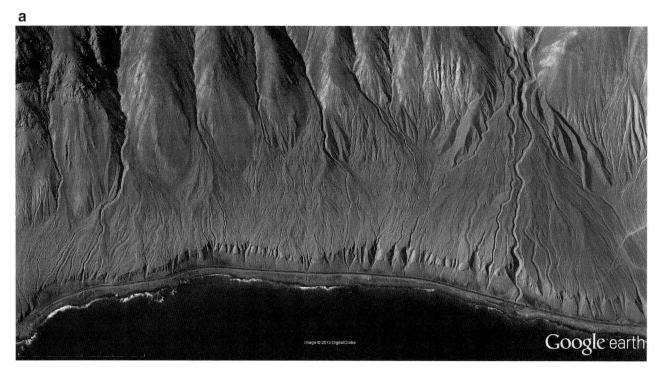

Fig. 8.11 (a–f) Large alluvial fans in northern Chile, stretching westwards from the coastal cordillera to the coast. Large parts of the fans are built up by debris flow deposits deriving from the slopes of the Cordillera. Pathways of debris flows with typical ravines and accompanying ridges as well as terminal lobes are particularly visible in (b–f) (Image credit: ©Google earth 2012)

8.2 Mass Movement of Unconsolidated Materials 177

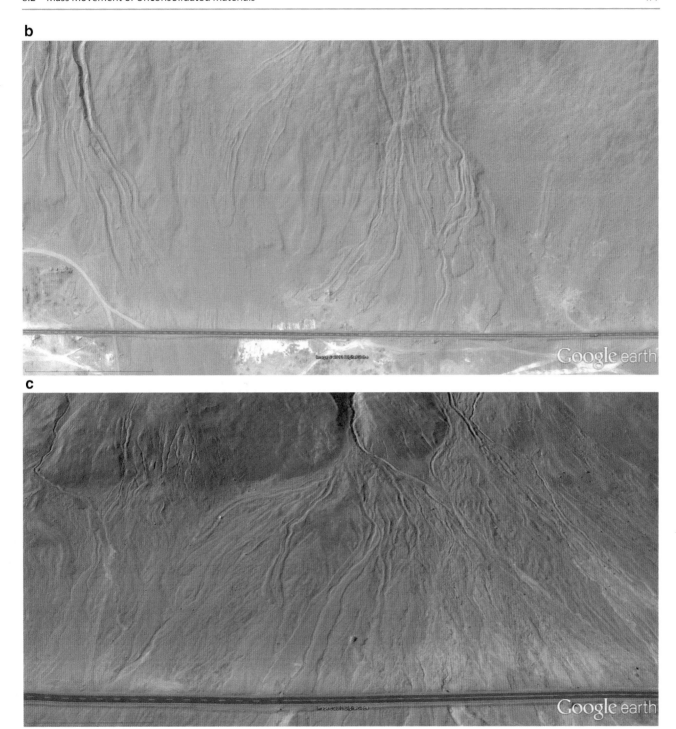

Fig. 8.11 (continued)

178   8 Mass Movements: Landforms Shaped Under the Force of Gravity

d

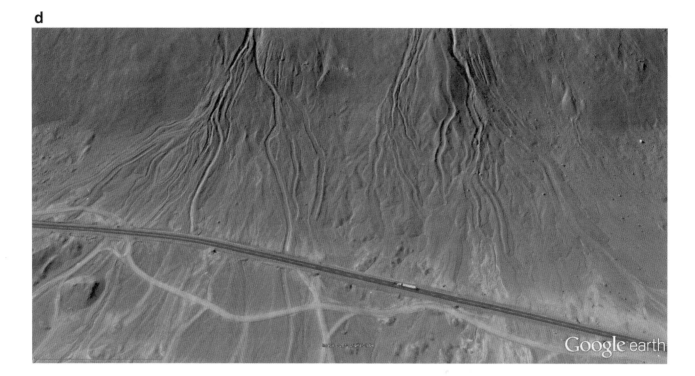

e

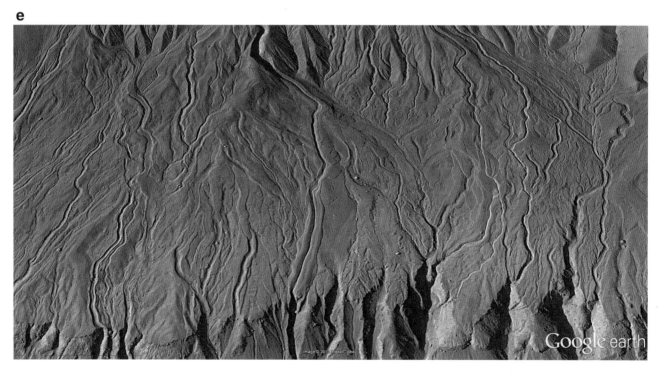

Fig. 8.11 (continued)

Fig. 8.11 (continued)

Fig. 8.12 La Casita landslide and lahar pathway in Nicaragua. On October 30th, 1998, hurricane Mitch resulted in intense rainfall in Central America, causing parts of the southern flank of the Casita volcano to slide. The original mass of the slide successively increased by incorporating volcanic material and water and turned into a lahar (from a hyper-concentrated flow and later a debris flow). The La Casita event exemplifies the dynamic complexity of numerous mass movement events, which are often composed of multiples stages with different characteristics of the process of movement. (see also URL: http://volcanoes.usgs.gov/hazards/lahar/casita.php) (Image credit: ©Google earth 2012)

Fig. 8.13 (**a**) Landslide scar on a steep slope on the Brazilian coast (*23°06′45″S, 44°14′59″E*). In January 2011, a period of heavy rainfall over several days caused the 2011 landslide disaster in the state of Rio de Janeiro, Brazil. The intense rainfall caused thousands of mass movements, in particular along the Serra do Mar between the states of Espirito Santo and Santa Catarina. The deep weathering of the crystalline basement favors the occurrence of debris slides, generating disintegrated rock debris and saprolite on the slopes directly above the crystalline basement. (**b**) The municipality of Nova Friburgo was heavily affected by the event. The image shows a rural area with numerous debris slides (*22°14′0.33″S, 42°38′7.49″W*)

Fig. 8.14 (a, b) The image sequence shows the same rural area in the municipality of Nova Friburgo (Brazil) before and after the January 2011 landslide event (*22°15′03″S, 42°34′54″W*) (Image credit: ©Google earth 2012)

Further Readings

Abbott PL (2004) Natural disasters. McGraw-Hill Companies Inc., New York

Antoniadis J, Freire PCC, Wex N et al (2013) A massive pulsar in a compact relativistic binary. Science 340(6131):1233232. doi:10.1126/science.1233232

Augenstein C (2007) Spuren des Flimser Bergsturzes im Dachlisee (Obersaxen). Jahresberichte der Naturforschenden Gesellschaft Graubünden 114:43–57

Eberhardt E, Stead D, Coggan JS (2004) Numerical analysis of initiation and progressive failure in natural rock slopes – the 1991 Randa rockslides. Int J Rock Mech Min Sci 41:69–87

Erismann TH, Abele G (2001) Dynamic of rockslides and rockfalls. Springer, Dordrecht

Evans SG, DeGraff JV (2002) Catastrophic landslides: effects, occurrence, and mechanisms. Geological Society of America, Boulder

Glade T, Anderson MG, Crozier MJ (2006) Landslide hazards and risk. Wiley, Southern Gate

Goudie AS (ed) (2006) Encyclopedia of geomorphology. Taylor and Francis, London

Humlum O (2000) The geomorphic significance of rock glaciers: estimates of rock glacier debris volumes and headwall recession rates in W Greenland. Geomorphology 35:41–67

Ivy-Ochs S, Heuberger H, Kubik PW, Kerschner H, Bonani G, Schluchterer C (1998) The age of the Köfels event. Relative, 14C and cosmogenic isotope dating of an early Holocene landslide in the central Alps (Tyrol, Austria). Zeitschrift für Gletscherkunde und Glazialgeologie 34:57–70

Kusky TM (2003) Geologic hazards. A sourcebook, Oryx sourcebooks on hazards & disasters. ABC-Clio, Santa Barbara

Margottino C, Canuti P, Sassa K (eds) (2013) Landslide science and practice, vol 4, Global environmental change. Springer, Dordrecht

NASA Earth Observatory. URL: http://earthobservatory.nasa.gov/Features/GRACE/page3.php. Accessed 8 Nov 2013

Petley DN, Bulmer MH, Murphy W (2002) Patterns of movements in rotational and translational landslides. Geology 30:719–722

Poschinger A, Wassmer P, Maisch M (2006) The Flims rockslide: history of interpretation and new insights. In: Evans SG (ed) Landslides from massive rock failure, NATO Science Series. Kluwer Academic, Dordrecht, pp 329–356

Prager C, Zangerl C, Patzelt G, Brandner R (2008) Age distribution of fossil landslides in the Tyrol (Austria) and its surrounding areas. Nat Hazards Earth Syst Sci 8:377–407

Stoffel M, Marston RE (eds) (2013) Mountain and hillslope geomorphology. In: Shroder J (editor in chief) Treatise on geomorphology, vol 7. Academic, San Diego

Forms by Flowing Water (Fluvial Features) 9

Abstract

Water is the most precious element on our planet, and therefore rivers also are called "veins of life", particularly in arid or semi-arid landscapes. Rivers, however, are the forces which form valleys by cutting linear to oscillating depressions into the landscape. Even if the river is no longer present, we can easily identify the valleys as the result of previously flowing water (e.g., by well-rounded and stratified sediments, and the typical geomorphology of valleys). Evidently, to cut valleys in hard rock requires more than water. Debris of different sizes (from silt to boulders) transported by flowing water are the real tools of erosion and are responsible for the incision of valleys. Valley incision, however, needs some time, and valleys therefore may be rather old forms, up to many million years old. Exceptions are gully-like forms in soft rock, which may start to form within a single strong rainfall event. All in all, valleys of different length, depth, or form and shape are amongst the most frequent features in geomorphology, which gives apparent evidence that our Earth is a "water planet" even on the continents. In this chapter we will present examples of valleys in different climates and topographies, which exhibit different conditions of flowing water and sediment transport. The first part deals with flowing water and the incision of valleys, with additional specific features like meandering or anabranching rivers. The second part shows how rivers accumulate sediments in the forms of debris fans or river terraces. River terraces (also fluvial or alluvial terraces) accompany most rivers of the Earth as elongated and level former floodplains above the present one. These terraces are relicts of fluvial sediment accumulation and/or fluvial erosion and give evidence of the long and complex history of a river's evolution indicated by periods of cut (fluvial incision) and fill (fluvial accumulation). As such, they indicate past changes of water discharge, stream power and sediment input, which may be triggered by climatic changes. In the temperate regions, generally known as areas influenced by strong climatic fluctuations during the last Ice Age, older valleys are filled by river sediments and the successive incision of the rivers into these sediments has formed a series of terraces along many rivers. In addition to climate, tectonics may influence terrace building by forcing the river to incise or deposit their sediments. Another form of fluvial deposition (except of fluvial deltas along the coasts of the Earth's oceans and along the shores of lakes) is alluvial fans. They usually have the form of cones, starting at the point where rivers abruptly exit steeper and narrow valley sections and enter lowlands with reduced inclination, distributing their debris to all sides in front of a mountain range.

Three processes are the most important for relief formation on Earth. These are flowing water, wind and moving glacier ice. However, water, wind and ice work the most effectively if weathering has prepared the surface and delivered debris as tools for carving. These natural elements alone – without using debris as tools – are much less efficient. This chapter deals with processes in flowing water and the resulting terrestrial landforms.

Our planet has many hundred millions of rivers (streams, streamlets, creeks, gullies etc.) (Fig. 9.1a–d), as well as an extremely high number of valleys without flowing water but formed by rivers for instance during different climatic conditions in the past. Valleys and rivers of considerable size (length, depth, and width) belong to rather old forms on Earth, and some valleys have formed during the last several million years. However, smaller valleys like gullies or those found in badlands which have formed in soft rock or unconsolidated deposits may be younger (Fig. 9.2a–e). The depth of a valley is not the best criterion to judge on its age, because this depends on the amount of water, the number of extreme flood events, the resistance of the underlying rock, its tectonic setting, and most importantly, the particles which have been transported during the time of formation. A better indicator for the age of a valley or a river system is its length, because valleys develop from low-lying areas (e.g., from the river mouth into a lake or the ocean or from the confluence with a larger river in its floodplain) backwards and into higher relief. It is a slow process potentially traversing across different petrologic units, topography and climatic zones.

Rivers often are called "veins of life", because they deliver water and food (fish, crabs, mussels) and represent routes of transport. Many rivers with changing runoff and regular floods deposit fine particles in their low-lying floodplains as overbank deposits, which are important in keeping agricultural fields fertile in many areas. In the past, rivers often have been venerated as gods due to their positive attributes. This, however, does not necessarily depend on the sheer size of a river, as we can see by comparing the largest in the world. Rivers like the Amazon, Congo, Parana, Orinoco, Mississippi-Missouri, Sambezi and Wolga did not have the exceptional importance in human and social evolution compared to the Nile, Indus, or Euphrates and Tigris in Mesopotamia. These four rivers cross wide arid landscapes, where water is the most precious commodity. Here, water has been used for irrigation even in very early times and periodic flooding of agricultural land guaranteed rich harvests (Fig. 9.3a, b). That is why the oldest civilizations on Earth developed along these rivers: the Egyptians with their pyramids from about 5,000 before present, or in Mesopotamia with Euphrates and Tigris with a similar age of civilization and the oldest large town of Ur. Writing and beer brewing was invented more than 4,000 years ago in Ur, when Abraham lived here according to the Old Testament of the Bible. The Indus culture has an age of nearly 5,000 years as well; so far more than 1,500 settlements have been found along the river, and this early culture covered an area of about 1.25 million km^2, which is more than that of old Egypt and Mesopotamia combined.

There are many more reasons to investigate rivers and valleys from different scientific disciplines, but our aim is to explain and illustrate some of the most important landforms which are caused by flowing water. Water in a river may flow in different ways from a physical point of view. If the flow is extremely slow as it appears in some large rivers near their mouth, a quasi-laminar motion can be observed, in which case the water particles move nearly parallel to each other without turbulent mixing. Laminar flow conditions, however, rarely occur in natural streams. In general, water flowing in a river is turbulent with a complex mixture of rollers or cylinders. In the upper and steeper sections of river valleys this turbulence is stronger with the incorporation of air bubbles and the rough surfaces of the flowing water. The character of flowing water depends on its velocity, which also depends on friction. Friction in a river can be observed along the river bed and along the river banks (and even at the surface in contact with the atmosphere and with the wind). Flowing is only possible if this friction is overwhelmed by the cylinders and rollers. During high water the zone of friction affects a smaller part of the flow profile, allowing the flow rate to increase while the incline of the river remains the same.

A general question is, from where and by which processes does a river gets its water? Normal rivers cannot be fed by springs alone (if there is one!). The water may come from rain directly into the river, but by far the largest portion of the water derives from surface runoff and groundwater flow from the surrounding landscape and by smaller tributaries, which depend on those conditions. A river may even flow after weeks without rain if it is fed by groundwater seeping through the river banks. During high water the river itself may feed the groundwater along and through its banks again.

In general, the ability of a river to incise and form a valley depends on the relation of its discharge volume and stream power, its flow characteristics (including the inclination along the longitudinal profile), the characteristics of the geological basement (setting, permeability, resistance to mechanical and chemical destruction), and the surrounding climatic conditions. The lattermost is important for the flow regime, but also for delivering the tools for erosion. These tools may be sand of different sizes or coarse rock fragments such as typical river pebbles, which may have been weathered from the hard rock by frost shattering in high mountainous or polar regions and then rounded during long periods of transport. If the river is able to transport these particles, preferably as bed load in saltation, the erosional effect will be greatest. Potholes form where cylinders develop repeatedly at the same place. If initial pools are deep enough, larger fragments may fall in and get trapped. They cannot escape,

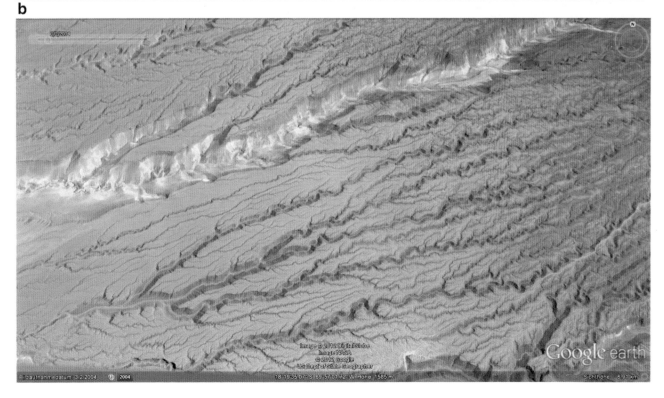

Fig. 9.1 (a) Landscape in eastern China dominated by fluvial processes. Countless creeks and smaller streams as well as few larger rivers have shaped the landscape. Scene is 55 km wide at *38°17′N* and *111°21′E*. (**b**, **c**) In the desert of northernmost Chile, the large and small valleys on the same geological basement have incised with similar swinging forms of different dimensions. Width of the landscape shown here is about 6 km at *18°19′S* and *69°57′W*. (**d**) Dendritic organization of valley systems in Morocco at *27°40′N* and *1°42′E*. The scene is 20 km wide (Image credit: ©Google earth 2012)

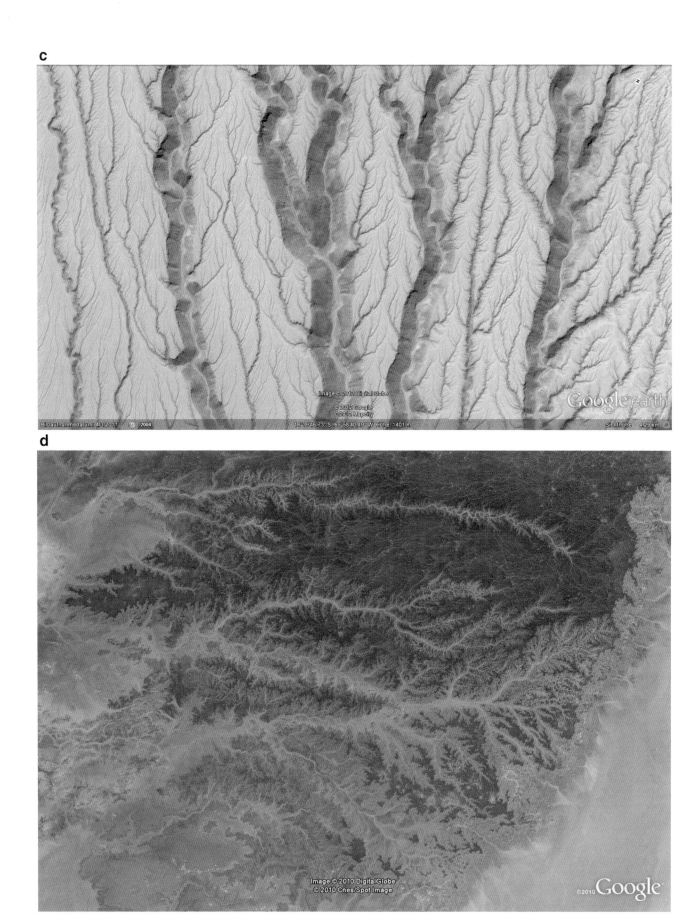

Fig. 9.1 (continued)

Fig. 9.2 (**a**, **b**) Badlands in soft siltstones at Zabriskie Point in Death Valley National Park, California (USA) formed during strong but rare torrential rains from summer thunderstorms. A black dike of hard basalt crosses the dissected landscape (Image credit: ©Google earth 2012/D. Kelletat). (**c**) In northwest Argentina, in an arid environment in the rain shadow of the Andes, the foothills consist of less resistant sedimentary rocks. Episodic runoff dissects formerly smooth hills into sharply incised badlands. Scene is 10 km wide at about *27°43′S* and *67°50′54″W*. (**d**) Eastern China loess landscapes are densely dissected by gullies to form a "badland" *sensu stricto*. Scene is 23 km wide at *35°38′N* and *105°55′E*. (**e**) Another aspect of the badlands in eastern China's loess areas, 55 km wide at *37°49′N* and *109°47′E* (Image credit: ©Google earth 2012)

c

Fig. 9.2 (continued)

d

Fig. 9.2 (continued)

e

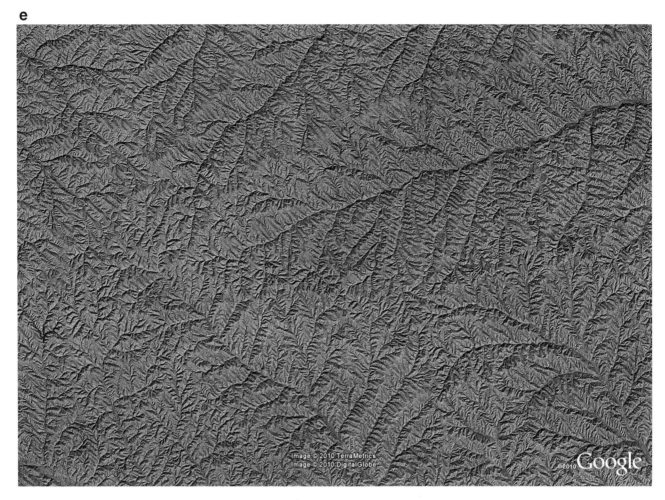

Fig. 9.2 (continued)

9 Forms by Flowing Water (Fluvial Features)

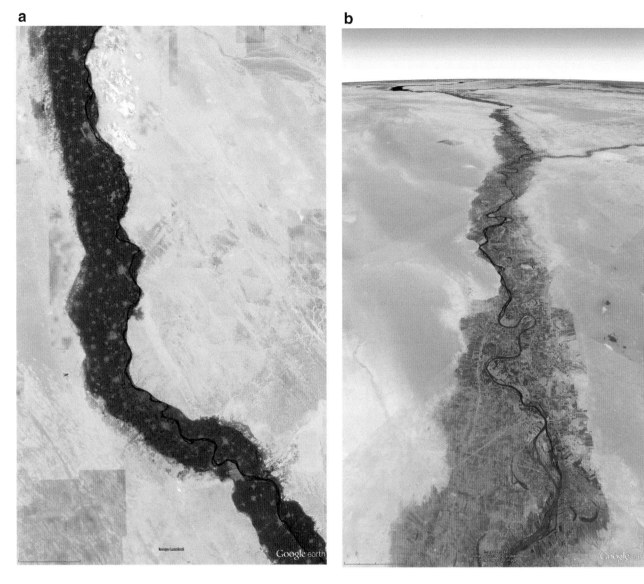

Fig. 9.3 Rivers like the Nile (**a**), the Indus, or the Euphrates (**b**) and the Tigris in Mesopotamia all flow across wide arid landscapes, where water is the most precious commodity. The river water has been used in the broad floodplains to irrigate the fields for millennia, and periodic flooding of agricultural land guaranteed rich harvests (Image credit: ©Google earth 2012)

Fig. 9.4 A deep pothole is filled with pebbles as erosive tools in a small river of central Alaska. During higher floods the potholes are submerged under flowing water and standing whirlpools develop within the shafts. Over time, pebbles and boulders will be eroded themselves; if no more pebbles are trapped in the pothole, the process of pothole formation will cease (Image credit: D. Kelletat)

but can move around in a whirlpool and enlarge and deepen the pothole (Fig. 9.4). Sometimes one large fragment will be enough to further erode the pothole, but it can also be worn down and at last pushed out, and fresh fragments may replace it. Smaller particles (such as silt) transported in suspension may be responsible for fine polishing the rock surface. If no "erosive tools" are present, incision will be minimal. In addition, sediment overload in rivers may hinder river bed erosion and the incision of a river. If the sediment load exceeds the transport capacity of a river, the sediment cannot be transported downstream and it will be deposited on the valley bottom, thus preventing the river bed from erosion.

Rivers ultimately transport particles from the continents into the oceans. Three different modes of transport can be distinguished. Transport by *solution* is invisible in most cases, but may become visible where huge amounts of organic substances are dissolved in the river water in areas with dense vegetation or peat. These are the so-called "blackwater rivers" in the tropics, owing to the dark color of the dissolved organics. The second mode of transport is *suspension* (Fig. 9.5a, b), which means that smaller particles cannot sink down in the river water due to the turbulence of flowing water. As a consequence, the water is enriched with smaller particles such as clay and silt; however, the size of suspended particles depends on the flow velocity and turbulence of the river. Suspension load can be seen easily in meltwater streams near the mouth of glaciers, where the light color of the meltwater derives from tiny rock particles and is referred to as "glacier milk". Finally, sediments may be moved as *bed load*, that is, the transport of heavier and larger particles on the river bed.

The dominant transport mode differs from river to river and depends on the weathering processes and therefore from the climate of the region. In tropical areas with dominant chemical weathering, rock components are entirely disintegrated and elements are dissolved in the water. Therefore, the portion of dissolved substances is very high in rivers like the Amazon, which is the reason for its dark color. Glaciers polish rock surfaces and produce fine particles (such as silt) which is best transported in suspension. Suspension transport can be seen in most rivers when high water discharge after strong rainfall events washes huge amounts of finer

Fig. 9.5 (**a, b**) Confluence of Rio Negro ("Black River") and the Amazon River west of Manaus, Brazil. The water from the Amazon main stream (and a number of its tributaries) comes from the Andes, as indicated by a light-colored suspension load. Solution is the more important transport mode in the lowland tributary Rio Negro and it carries dissolved organic substances, all dark in color. For this reason, most lowland tributaries of the Amazon are called "black water rivers" Considerable differences in the density of the two waters hinder a mixing of the waters for many kilometers downstream (Image credit: ©Google earth 2012)

Fig. 9.6 The Hjulström diagram. It shows the relationship between the grain size transported and the flow velocity in a river. It seems logical that the faster the flow, the larger the particles which can be moved. But we also see that with very tiny particles, increased flow velocity is needed to set them into motion (transport) and they can remain in suspension even if the flow velocity decreases (Image credit: A. Hager)

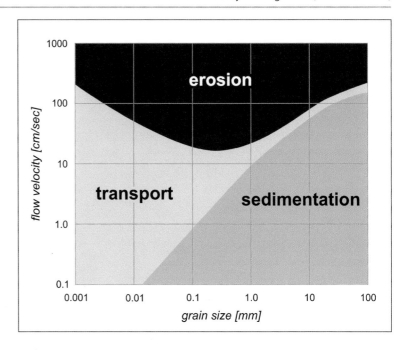

particles from the surrounding slopes and soils into the channel. In high mountainous areas with steep inclination of river beds, the rock fragments produced from frost weathering are often too coarse to stay in suspension, and therefore they roll, slide or bounce (saltation) as bed load. Of course, all transport modes may occur contemporaneously and at the same place.

The relationship between flow velocity and the size of particles moved was studied by the Swedish scientist Hjulström some 80 years ago (Fig. 9.6). To understand the diagram we have to know how a river will pick up a particle in its river bed. This is not by pushing it forward, because the vector of this push is directed against the bed and stronger friction will arise. In contrast, the motion is initiated by an area of lower pressure at the leeward side of a particle, pulling the fragment upwards into the flow. The low pressure situation, however, can only work if the particle is large enough to allow differential pressure conditions around it. If the particles are very small as silt or clay, they form a smooth bottom layer without any chance of pressure differentiation. Together with the increased cohesive forces of fine particles such as clay, this is the reason why it needs a higher flow velocity for initiating the transport of fine silt and clay compared to sand. If by chance (e.g., if we disturb the system with a stick) these smaller particles are set into motion, we will see that they can now be transported easily a long way in suspension. Thus, large and very small particles need a high flow velocity to be moved, but once moved, the fine ones will remain in movement even if the flow velocity decreases again.

We may distinguish rivers based on their runoff characteristics throughout the year. Permanent or perennial rivers discharge water around the year and may be found in the humid and semi-humid tropics as well as areas in the mid-latitudes. Periodically flowing rivers occur in latitudes with distinct wet (discharge) and dry (no discharge) seasons (Fig. 9.7a, b). However, periodicity occurs also in very cold regions, where flowing water is restricted to a short melting phase in late spring and 95 % of the yearly discharge runs off within a few weeks, and the river is frozen to the ground for the rest of the year. Finally, episodically flowing rivers can be found in dry climates and arid or semi-arid environments (deserts), where there is enough water to fill and activate a river bed only during very rare (sometimes extreme) events of precipitation that may occur every decade or so, or even less frequent (Fig. 9.8a, b). Nevertheless, from a geomorphological point of view, these episodic flows such as flash floods may be extremely effective, since a lot of particles are produced during inactive periods and are used by the flash flood as erosion agents. There are also valley systems with incisions deep into the geologic basement, but with no water flows at present. They may provide evidence that in former times the regional climate was different (Fig. 9.8a, b), or that the course of the former river has changed upstream.

The shape of river valleys may exhibit remarkable variations. Rivers thus may exhibit a number of different cross sections, depending on the relation between downwards directed fluvial erosion and denudation processes along the valley slopes. Very shallow trough-like valley shapes are common in semi-humid and plain landscapes such as eastern Africa, while V-shaped valleys of different widths are typically found in low mountain ranges of present mid-latitude climates with permanent (perennial) rivers (Fig. 9.9a, b). Cross sections with vertical walls or those stepped in adaption to rock resistance are called gorges and canyons (such as

9 Forms by Flowing Water (Fluvial Features)

Fig. 9.7 Typical periodic river in western Morocco, east of Agadir, a tributary of the Oued Sous (*30°31′N, 9°03′W*). The main channels of the braided river bed are filled with flowing water in (**a**) (March 24th, 2009), but the channels are dry in (**b**) (May 10th, 2009). Car tracks can be seen where the river bed becomes dirt roads to cross the channel. Scale is 260 m long (Image credit: ©Google earth 2012)

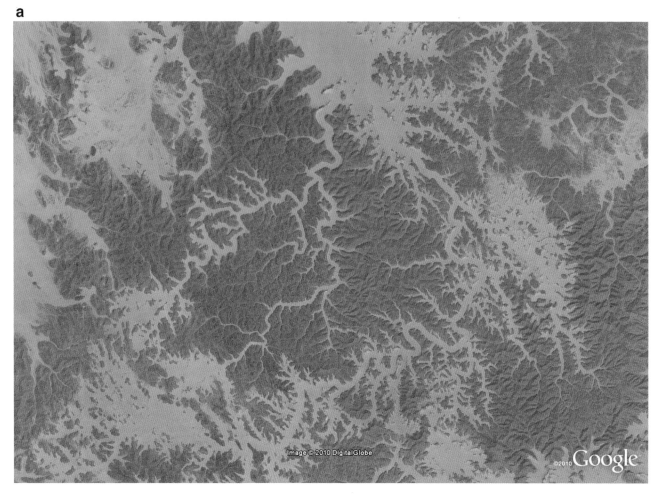

Fig. 9.8 (a) A meandering river valley in dry landscape, southernmost Saudi Arabia (about *19°51′N* and *44°09′E*) shows well developed meandering river beds dissecting the mountains in between. They probably derive from a much more humid climatic period. Scene is about 20 km wide. (b) In northwestern Saudi Arabia, the Nefud desert exhibits some old (now dry) river courses, seen by the meanders from west to east in this ~25 km wide image at *28°33′N* and *35°55′E* (Image credit: ©Google earth 2012)

b

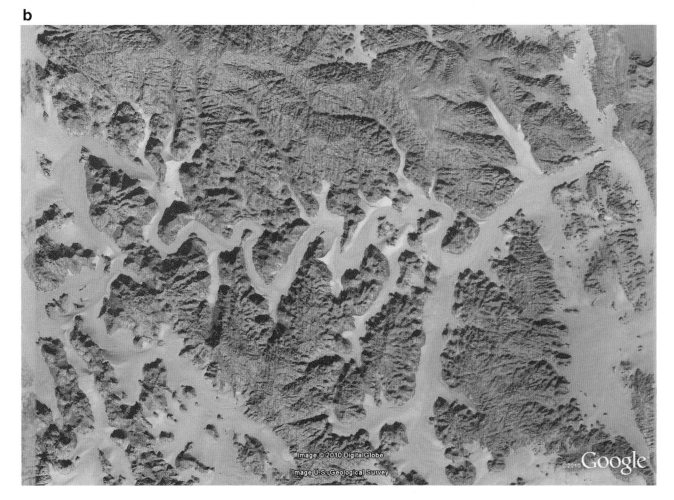

Fig. 9.8 (continued)

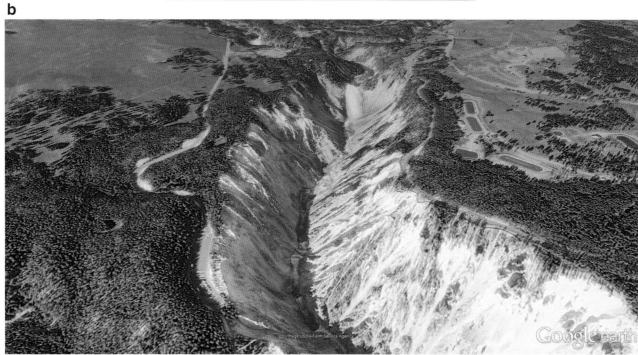

Fig. 9.9 (a) V-shaped valley cross profile, about 250 m deep, with a 60 m high waterfall of the Yellowstone River in Wyoming, USA (Image credit: D. Kelletat). (b) The same Yellowstone River section as seen from Google Earth (*44°43′N, 110°28′W*). The V-shaped valley form is clearly visible from space (Image credit: ©Google earth 2012)

the Grand Canyon in Arizona, USA), respectively (Figs. 9.10a, b and 9.11a, b). Canyons mostly form in arid landscapes with horizontally lying rock layers (such as less resistant sandstones), and gorges in steep and high mountains (Fig. 9.12a, b).

We rarely find a river or valley on Earth which is completely straight along its entire course. The typical swinging of rivers is called meandering, and each bend is called a meander, deriving from the western Anatolian rivers with the Turkish word *menderes*. Swinging of the flowing water may be initiated just by an irregularity on the river bed, or just a crosswind. The flow then swings to one side of the river, and by inertia it attacks this river bank more than the opposite bank. This results in high flow velocities and erosion at the attacked bank (undercut slopes) and low flow velocities and accumulation (point bars at slip-off slope) at the opposite inner bank (Fig. 9.13a, b). Further downstream, it will swing back to the other side of the river attacking the opposite bank. This swinging may continue for a very long passage of the river, sometimes along its entire course (Fig. 9.14a, b). The width and the curvatures of the meanders appear to be rather similar, but in fact they normally show a ratio of 1:2 or 1:3. The ongoing meandering of rivers with constant erosion at undercut slopes and deposition of point bars at inner banks (slip-off slopes) causes ongoing changes of the river channel course, including the migration of single meanders (Fig. 9.15). Where undercut slopes migrate towards and finally connect to an adjacent meander, the flowing water will use the new shorter channel, and the river will abandon the old meander creating a dead meander (Fig. 9.16a, b). Thus, this process is responsible for the formation of so called oxbow lakes (disconnected dead meanders; Fig. 9.17) and narrow necks (the narrow banks remaining between two opposite undercut slopes). At some point, a first exceptional flood event will overflow the narrow neck and consecutive floods will easily overflow and lower the neck, leading to its final breakthrough; the river will then take the shorter course, and the former bend will be abandoned. However, swinging and bending can also be controlled by structural reasons such as changes in rock resistance, dipping of strata in different directions, or crossing of fault lines (Fig. 9.18a–c). The following images give further examples from the varieties nature presents to our eyes (Fig. 9.19a–c).

So far we have discussed the meandering of a river, and the images mostly showed examples where the river was free to flow in its flood plain, developing free meanders. Now we will have a look at the meandering phenomenon, in which rivers that are constrained in sometimes deep valleys cut into the hard rock and swing in similar forms (Fig. 9.20a, b). The only explanation for meandering valleys is that they derive from rivers with free meanders on a former plain as predecessors, and that during tectonic uplift incision of these meanders started and the forms have been preserved for the time of incision. Of course changes in the geomorphology of the river valley will appear during this long period of time (hundreds of thousands of years as a mean order), and still lateral erosion will steepen undercut slopes and alter the course of the river. The following images may give some impressions of valley meanders and their most significant details (Figs. 9.21a–c and 9.22a–c).

We also find rivers or river sections where sediment deposition is dominant. Compared to the river discharge, the amount of debris in these cases exceeds the transport capacity of the river and the accumulated sediment in the river bed becomes obstacles for the flowing process itself. In these cases, the river will deposit a certain portion of the sediment forming elongated sand or pebble bars. The river will split its course and flow along both sides of these bars, mostly forming "islands" of sand, pebbles or cobbles. When this process leads to many channel arms in the same valley cross section, and many bars in between, we use the term "braided river" (Figs. 9.23a–d and 9.24a–c). The sand and pebble bars of braided rivers are typically ephemeral, unconsolidated, and poorly vegetated. Once the discharge of the river and stream power increase, the sediment is entrained by the flowing water again and the bars may alter or disappear and new bars may form in the river bed. Braided rivers are typical for permanent (perennial) rivers in areas with high rates of physical weathering, but they often occur in areas with periodic or only episodic flow as well, where discharge is reduced and weathering products are produced in high rates. Although downward directed erosion is reduced or absent in braided rivers, lateral erosion takes place at undercut slopes along the river banks. This leads to a cross profile of a trapezoid-like form: a flat bottom and more or less steep but mostly not very high banks.

Anabranching rivers are indicated by multiple channels or river arms as well. In contrast to braided rivers, anabranching rivers are characterized by vegetated and/or stable alluvial islands that divide flows (i.e., the different arms of the river channel) at discharges up to bankfull (Figs. 9.25a, b, 9.26a, b, and 9.27). Finally, anastomosing rivers represent a special sub-type of anabranching rivers. However, the transport capacity of a river may vary along its longitudinal profile, for instance due to changing inclinations along different river sections or varying sediment input. Therefore, different sections along the longitudinal profile (see below) of a river may be described as braiding, anabranching, and meandering.

Lowland rivers with a significant suspension load during high water may build up dams along their banks, often of the same dimensions along long river sections. The dimensions of these dams depend on the discharge and suspension load of the rivers. During overbank floods, this suspension load (silt and sand) is deposited directly beside the river bed on top of the banks, where vegetation and increased friction reduces the flow velocity and the sediments accumulate.

Fig. 9.10 (**a**) The Grand Canyon of the Colorado River in northern Arizona (USA) exhibits its approximately 1.6 km deep and 20 km wide cross section of steep and smoother slope sections in relations to the resistance of the rock strata to weathering and erosion. This is a typical stepped canyon profile (Image credit: S.M. May). (**b**) The Grand Canyon from Google Earth (Image credit: ©Google earth 2012)

Fig. 9.11 (a) An oblique aerial photograph of the Little Colorado River gorge cutting more than 350 m deep into a sequence of uplifted, near horizontal limestones and sandstones in northern Arizona, USA (Image credit: D. Kelletat). (b) The Little Colorado River gorge as seen from Google Earth (Image credit: ©Google earth 2012)

Fig. 9.12 (**a**, **b**) Gorge of the Aare River in Switzerland (*46°43′04″N, 8°12′E*). View is to the northwest in (**a**) (Image credit: ©Google earth 2012 and D. Kelletat). The gorge (and several further ones, hidden under glacial deposits) has been eroded since glacial conditions under a thick glacier, where meltwater is running under hydrostatic pressure

9 Forms by Flowing Water (Fluvial Features)

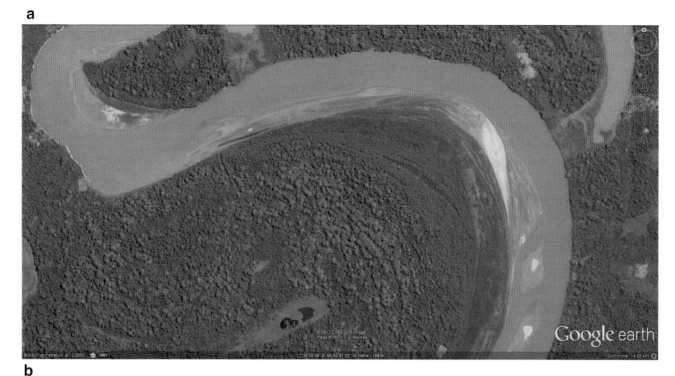

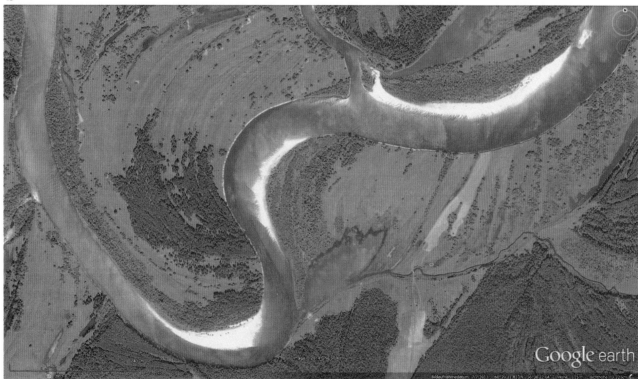

Fig. 9.13 (a) Detail of inner bank deposits with a perfect example of point bar sedimentation (Rio Beni, northern Bolivia, *11°39′S* and *66°42′W*). (b) Point bar sedimentation along the inner banks of the meandering Vishera River in Russia, west of the Ural Mountains (*60°20′N* and *56°38′E*). High flow velocities occur along the talweg (channel line) in the outer bank of the meander, where erosion creates undercut slopes (Image credit: ©Google earth 2012)

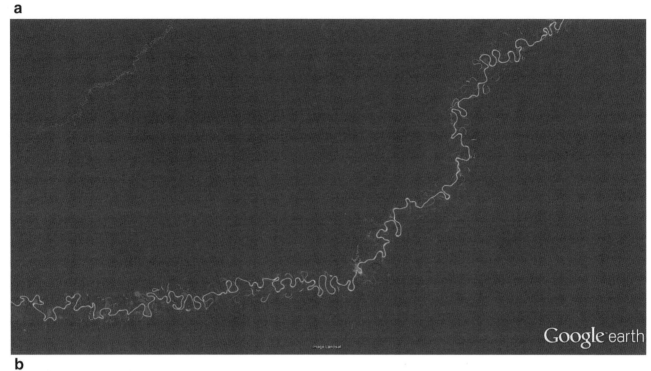
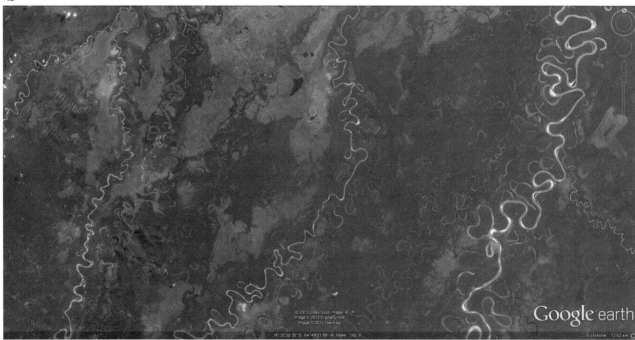

Fig. 9.14 In the tropical rainforest plains of (**a**) western Brazil (Rio Juruá, *5°59′S* and *67°57′W*), or (**b**) Bolivian lowlands tributaries (Rio Ichilo and Rio Chapare, *~16°29′S, 64°59′W*) of the Rio Mamorè. Here, the Amazon River system exhibits these perfect meanders and countless oxbow lakes. Sections are 270 km (**a**) and 80 km (**b**) wide (Image credit: ©Google earth 2012)

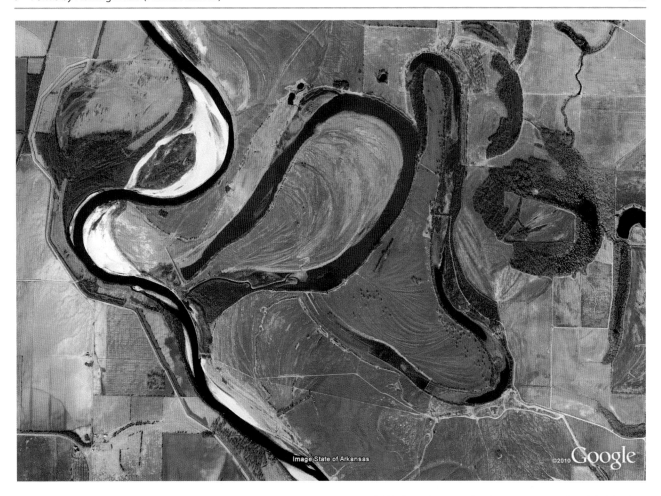

Fig. 9.15 Detail from the middle section of the Mississippi River at *33°25′N* and *93°43′W*. The meander shown (now an oxbow lake) is about 3 km wide (Image credit: ©Google earth 2012)

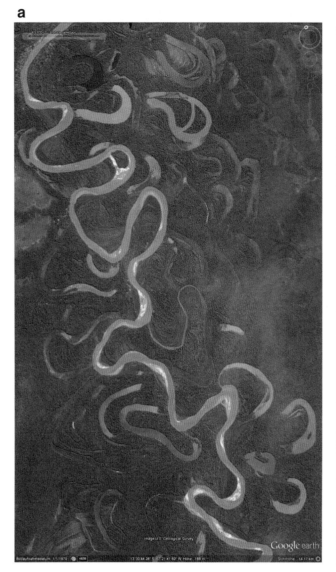 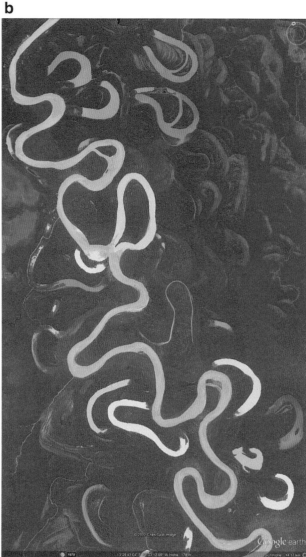

Fig. 9.16 Two images of the same river area (eastern Bolivia, Rio Beni, *13°25'S, 67°21'W*) taken in 1970 (**a**) and 2012 (**b**). The shifting of meander arms, the meander cut-off and the evolution of oxbow lakes is clearly visible, demonstrating the high geomorphological activity of fluvial dynamics. In the central part of image (**b**), the meander cut-off just occurred and the old meander arm will develop to an oxbow lake in the near future (Image credit: ©Google earth 2012)

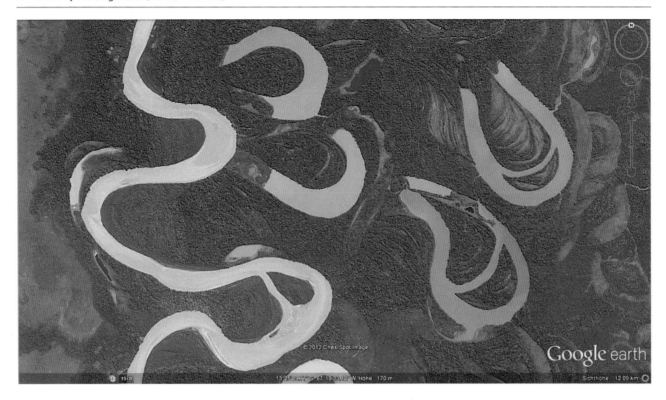

Fig. 9.17 Detail of oxbow lakes. Beside old meanders cut off from the main river, a lot of older contours of the river course can be seen in the eastern part of the image. Scene is taken from the same area as in Fig. 9.16, northern Bolivia (Rio Beni) (Image credit: ©Google earth 2012)

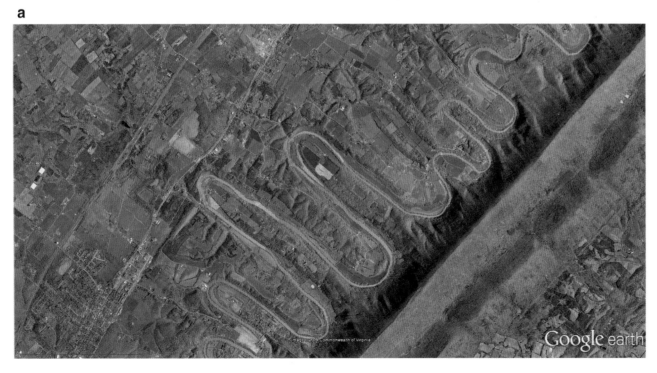

Fig. 9.18 River meanders mainly controlled by structural circumstances. (**a**) Meanders of the Shenandoah River, Appalachian Mountains at *38°54′N* and *78°28′W*, "caught" within the fold structures of the Appalachian Mountains. (**b**) Meandering with structural control (adapted to folds/cuestas) in the middle Colorado River (Nevada, USA) at *36°11′N* and *114° 02′W*. Scene is 9 km wide. (**c**) Structural control by two main joint directions of a meander in Marble Canyon of the Colorado River, west of Page (Arizona, USA) at *36°51′N* and *111°34′W*. Scene is 10 km wide (Image credit: ©Google earth 2012)

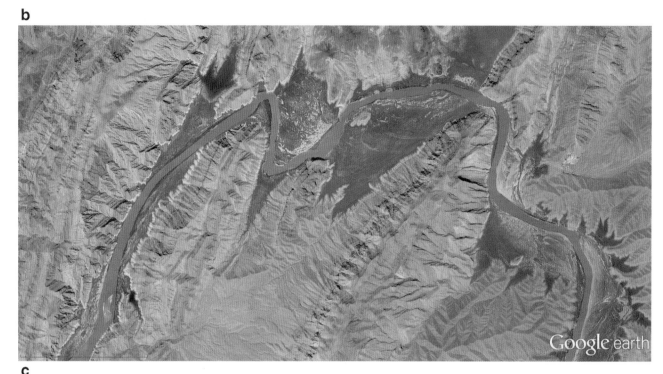

Fig. 9.18 (continued)

9 Forms by Flowing Water (Fluvial Features)

Fig. 9.19 (a) All rivers in their bends erode the outer bow (undercut slopes) and deposit sediments at the opposite side in the inner bow (inner bank, slip-off slope). The migration of the meander by erosion along the undercut slopes and sedimentation at the inner banks produces a sequence of several older inner bank deposits. Example is from northern Australia at *15°07′S* and *141°43′E*. If the outer parts of a meander come closer by lateral erosion of the banks, a breakthrough at the most narrow section may occur during a flood event. This shortens the flow way significantly, causing increased flow velocity, stream power, and erosive energy. The shorter way (through the former neck) will be adopted by the river, and the old meander will be abandoned to form an oxbow lake (see also Fig. 9.17). (b) The meandering Murray River, Australia's longest river, at the state border of South Australia, New South Wales and Victoria. (c) Fine meanders of the Shenandoah River in Virginia, USA between the cuestas of the Appalachian Mountains. The image is 15 km wide at *36°38′N* and *82°28′W* (Image credit: ©Google earth 2012)

c

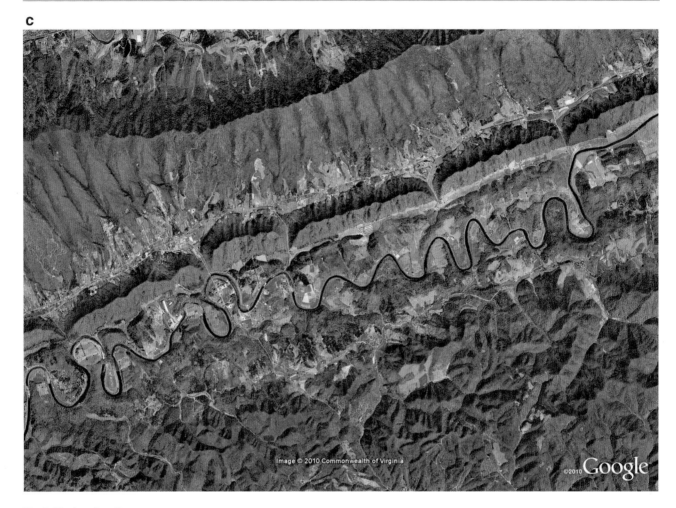

Fig. 9.19 (continued)

9 Forms by Flowing Water (Fluvial Features)

Fig. 9.20 (a) Valley meanders, 300 m deep, of the San Juan River in southern Utah, USA at about *37°11'N* and *109°55'W*. The river section shown is about 55 km long. The river crosses an anticline which has been uplifted a long time after the San Juan, the main tributary of the Colorado from the east, meandered on a plain (Image credit: ©Google earth 2012). (b) These meanders of the Finke River valley in central Australia near Alice Springs have been developed on a former floodplain as free river meanders. They have incised due to the successive unearthing of the anticline structure. This is a perfect example for a so-called "antecedence", where the river is older than the structure it cuts. Image shows same river section as in Fig. 4.9 (Image credit: D. Kelletat)

Fig. 9.20 (continued)

9 Forms by Flowing Water (Fluvial Features)

a

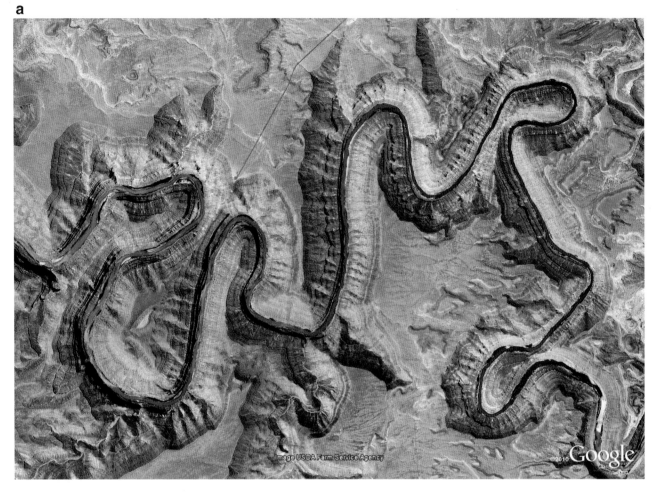

Fig. 9.21 (a) Valley meanders of the middle San Juan River in southern Utah (USA) at about *37°10′N* and *109°55′W*. This scene is 6 km wide, in which the river is at least 20 km long. (b) An extremely narrow neck of the San Juan river in its 250 m deep gorge at *38°36′N* and *110°02′W* (Image credit: ©Google earth 2012). (c) Goosenecks State Park of Utah, USA, near the small township of Mexican Hat. The river is the San Juan with an incision some 300 m deep (Image credit: D. Kelletat)

Fig. 9.21 (continued)

9 Forms by Flowing Water (Fluvial Features)

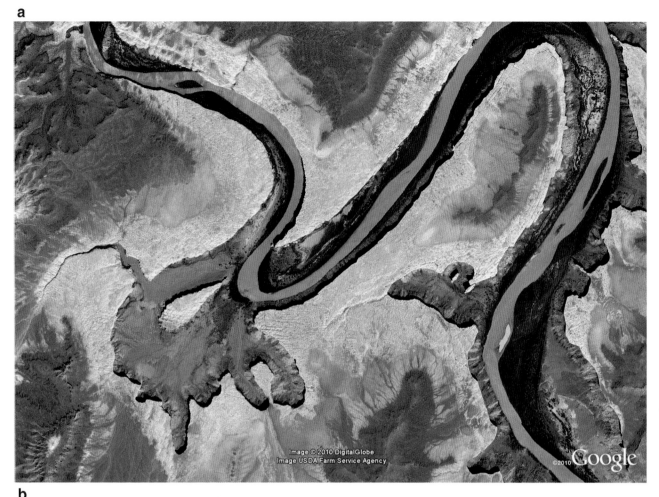

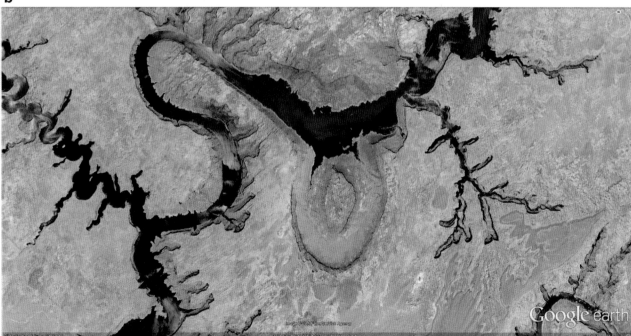

Fig. 9.22 Old meanders cut off from the river by a breakthrough of the neck a long time ago, when the river bed had a higher elevation. (**a**) Green River in Canyonlands National Park near Moab, Utah (USA) (**b**) Colorado River at *37°17′N, 110°48′W*. (**c**) The San Juan river before it reaches Lake Powell Reservoir near Page, Arizona (USA) (Image credit: ©Google earth 2012)

c

Fig. 9.22 (continued)

Fig. 9.23 (a) Glacial valley with broad braided river in New Zealand (*43°44′S* and *170°09′E*). A terminal moraine is located in the northeast and delivers huge amounts of rock debris material. (b) Another broad braided river in New Zealand (*43°36′S* and *170°29′E*). Rock debris is contribute to the valley by countless talus cones and debris fans, the lower parts of which can be seen to the east and west of the valley (Image credit: ©Google earth 2012). (c) Braided river in the lower Quebrada del Toro (Rio Toro), northwest Argentina. The photo was taken at ~*24°53′S, 65°40′W* looking to the west. Pebble bars are ephemeral and change during every flood (Image credit: J.-H. May). (d) Sand braided Penna River in southeast India. The sediment transported by river in this section is dominated by sand (*14°33′N* and *79°29′E*) (Image credit: ©Google earth 2012)

Fig. 9.23 (continued)

9 Forms by Flowing Water (Fluvial Features)

a

Fig. 9.24 (**a**) Meltwater outflows from glaciers may form very broad braided rivers as in this fluvial plain in southern Iceland, fed by the Vatnajökull (Vatna glacier), the largest of Europe (Image credit: D. Kelletat). (**b**, **c**) Broad braided river plains surrounding the Vatnajökull in southern Iceland (*63°51′N* and *16°59′W*) (Image credit: ©Google earth 2012)

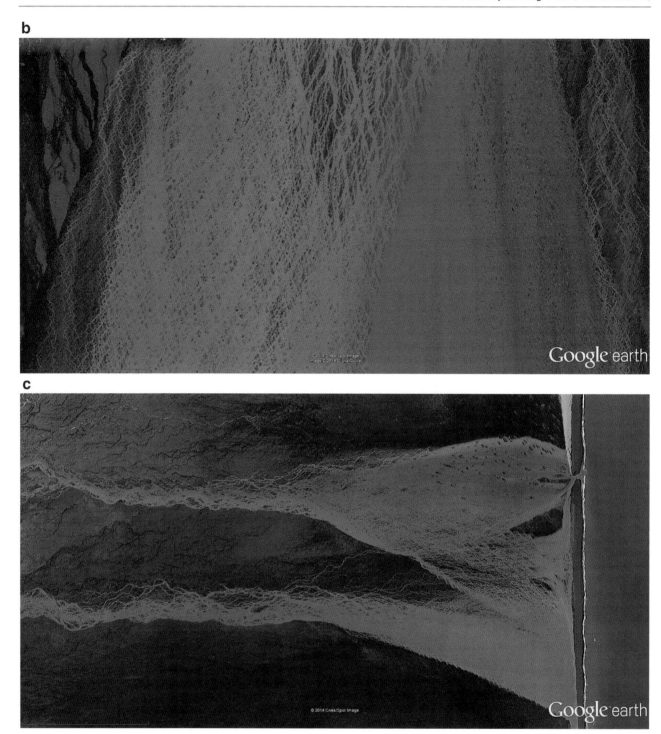

Fig. 9.24 (continued)

9 Forms by Flowing Water (Fluvial Features)

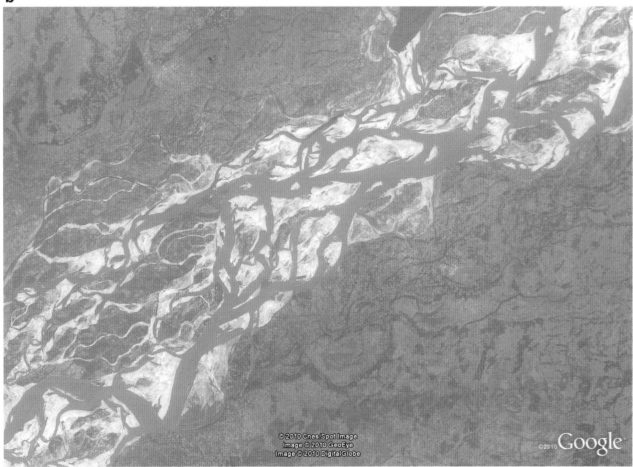

Fig. 9.25 (a) An anabranching river on New Zealand's south island at about *43°07′S* and *170°19′E*, showing a 2.6 km long section. However, both braided and anabranching sections of the river bed are present. (b) The middle course of the Brahmaputra River in northern India at about *26°25′N* and *92°12′E*. The river shows both braided and anastomosing character. Scene is about 10 km wide (Image credit: ©Google earth 2012)

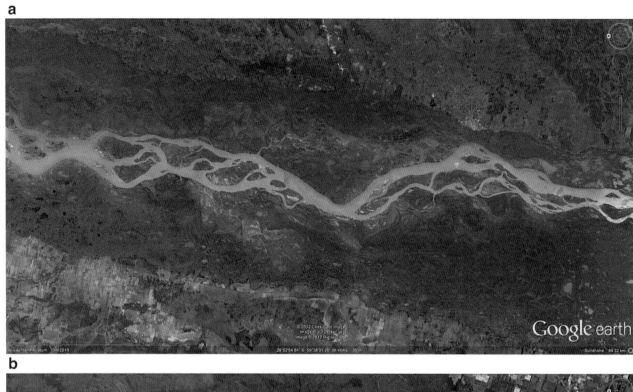

Fig. 9.26 (**a**, **b**) The Rio Parana in western Argentina is an excellent example for an anabranching river (*29°52′S, 59°38′W*) (Image credit: ©Google earth 2012)

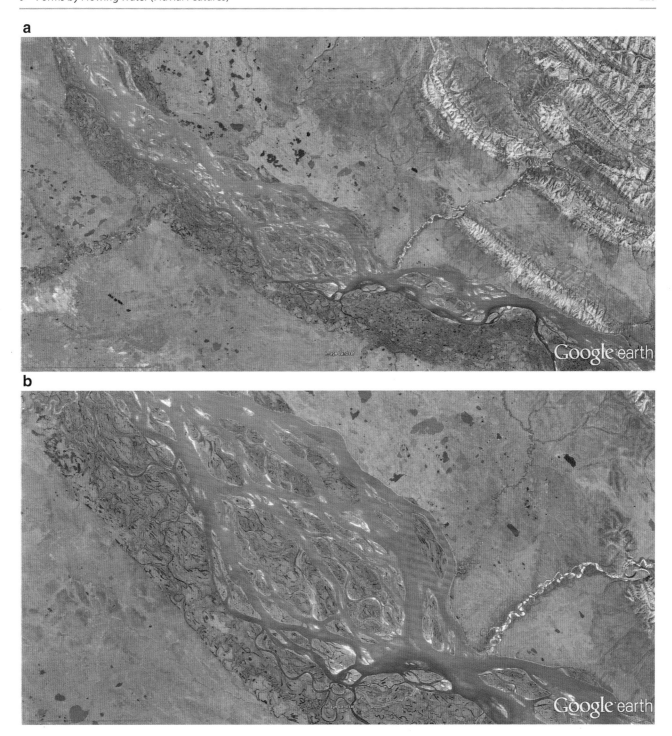

Fig. 9.27 (a) Perfect examples of anabranching rivers are found along the Lena River in Siberia, Russia, at about *64°43′N, 125°10′E*. (b) A detail image of (a) (Image credit: ©Google earth 2012)

More distant to the river banks the amount of sediment load is insufficient to accumulate in the same amount. The river thus forms broad (meters to kilometers) and low (decimeters to meters) natural embankments, so called "levees", accompanying a river for long sections. These natural levees are characteristic for lowland rivers or those flowing in a flat topography. Along the lower Mississippi River, they may reach a height of many meters and a width of several kilometers, slightly sloping away from the river. Where sedimentation additionally occurs in the river bed, the river will rise over the surrounding landscape, and the elevation of the river bed may become higher than the lowlands though still constrained by the two parallel natural levees. It is easy to imagine that catastrophic floods cutting the levees and overflowing into the adjacent low-lying plain will cause a dangerous and significant flooding. Flooding then may last for a long time, because the water flow is hindered by the dam-like natural levees and the water cannot flow back into the river channel. That is why people living along these rivers have to build additional high dikes on the levees for protection. Figures 9.28a, b and 9.29 show examples of natural levees.

Finally, the longitudinal profile of a river describes a topographical transect from its origin to its mouth. It may vary considerably between different rivers, and different sections of the longitudinal profile of one river may have considerably different characteristics. However, nearly all rivers and small creeks have an initial and higher section which is rather steep, whereas in the lowermost sections of a river's longitudinal profile, inclination is usually nearly invisible. An intermediate section may separate these two parts of the longitudinal profile (Fig. 9.30a–e).

In general, rivers tend to establish a longitudinal profile resembling the natural exponential function by compensating the varying topography along the river course by deposition and erosion; the uppermost steep section shows an incised valley (which means, it is erosive) and the lowermost part is dominated by sediment deposition. In the intermediate section, sediment transport (in suspension and bed load) without significant erosion or deposition may dominate. Since the uppermost section is generally erosive, the river tends to cut back into the landscape and elongate the valley backwards and upwards (Figs. 9.30a–d, 9.31a, b, and 9.32a–c). This back-cutting of inclining parts of a river is best seen at waterfalls, which also move upwards (and backwards). At Niagara Falls (Fig. 9.33a, b) the river has extended for more than 15 km since the last 10,000 years and is currently cutting back several decimeters per year. The falls will be transformed into rapids in the geologic future, and ultimately these rapids will be levelled by the tendency of the river to form a smooth, concave longitudinal profile.

The morphological process to form a deep valley needs a significant amount of time, in the order of hundreds of thousands of years or even several million years. During this long period of valley formation and river incision, rivers usually leave behind fill (depositional) or strath (erosional) terraces, reflecting the long history of valley evolution. Fill terraces are depositional bodies of fluvial sediments into which the river has incised its present bed. They formed at a time when the river was flowing at a higher elevation in the past, filling the valley with its sediments. When the river cut its channel down and created a new floodplain at a lower elevation, the remains of these former valley bottoms usually form elevated, elongated and level terraces on both sides of the river. Series of terraces may be left in the valleys by repeated periods of accumulation and incision of rivers, a process that has created the stepwise morphological pattern of many river valleys on Earth (Figs. 9.34, 9.35a–c, and 9.36a–c).

Regional uplift of the geological basement may promote incision of rivers and, consequently, the formation of terraces. However, in many regions of our Earth, terrace sequences are interpreted to be the morphological indication for past climates and climate changes. Generally speaking, long-term climatic changes may influence river discharge and stream power, weathering processes, and ultimately the sediment transport occurring in a river. River discharge may shift from a lot of water with only minor sediment, to less water with abundant sediment. In the first case, a period of dominant erosion may take place, and in the latter, a period of major deposition. The periods of dominant deposition lead to the aggradation of fluvial sediment in the valley, which may be cut again (e.g., during a more humid climatic period). One of the main reasons for terrace formation along high- to mid-latitude rivers is the multiple change from glacial (cold conditions that were dominant until ~12,000 years ago) to interglacial (warm conditions, compared to the climate of the Holocene starting ~12,000 years ago) periods during the Quaternary (i.e., the last 2.4 million years). Generally speaking, rivers had to face mostly cool conditions and cold winters during glacial periods, where frost shattering produced huge amounts of rock debris that were transported from the slopes to the river by solifluction or gelifluction. When ice and snow melted during the cool summer, the debris exceeded the transport capacity of the river, consequently leading to braided river conditions. Conversely, in interglacial periods, humid and warmer climatic conditions allowed discharge all year long, and frost shattering and slope processes were reduced. Rivers thus had to handle less debris with more water available, enabling it to incise into its own former bed. It is, in fact, assumed that the geomorphological processes of deposition (accumulation of material due to braided river conditions) and erosion (incision into its own deposits) were most efficient in the transitional periods when climate fluctuated from warm to cold conditions and *vice versa*.

A further important type of fluvial deposition is represented by debris fans (alluvial fans). They usually form along

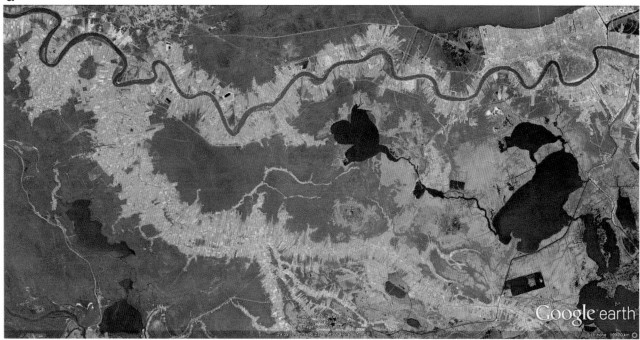

Fig. 9.28 (**a**) Along the lower Mississippi River (Alabama, USA), strips of cultivated land are present along the river on both sides; whereas further apart from the river, the swampy land is not cultivated. The reason is that only on the "natural levees" the ground is high enough to remain above flood water levels and therefore was used as the first places for settlements and cultivation during the time of French colonization (which still is documented by the special strip-like fields). Image in (**b**) is a detail image of (**a**) (Image credit: ©Google earth 2012)

Fig. 9.29 Natural levees along a small river near its mouth into Lake Titicaca in the Andean highlands of Peru at about *15°34′S* and *69°55′W*. The scene is 3 km wide. The Inca people first cultivated the natural levees, which is illustrated by the brown strip-like fields along the river course (Image credit: ©Google earth 2012)

Fig. 9.30 Differences in the geomorphological pattern along a river's longitudinal profile. (**a**) Upper section of the Amazonas catchment, Rio Apurimac, one of the headwater streams of the Amazon River. Steep slopes, rapids, and a V-shaped valley characterize this upper section of the river in the Andes. (**b**) Rio Ene, located further downstream after several confluences, at the foothills of the Andes. Here, the river is anabranching, caused by reduced inclination of the river bed and high sediment load. (**c**) Rio Ucayali, further downstream, after additional confluences. The inclination is further reduced, and the anabranching character passes into a meandering river. (**d**) The lower Rio Ucayali is indicated by its perfect meandering characteristics before reaching the Amazon. (**e**) Rio Ichilo and Rio Chapare entering the Bolivian lowlands (*~16°57′S, 65°11′W*). At the foot of the Andes we see a transition from anabranching to meandering. View is to north-northwest (Image credit: ©Google earth 2012)

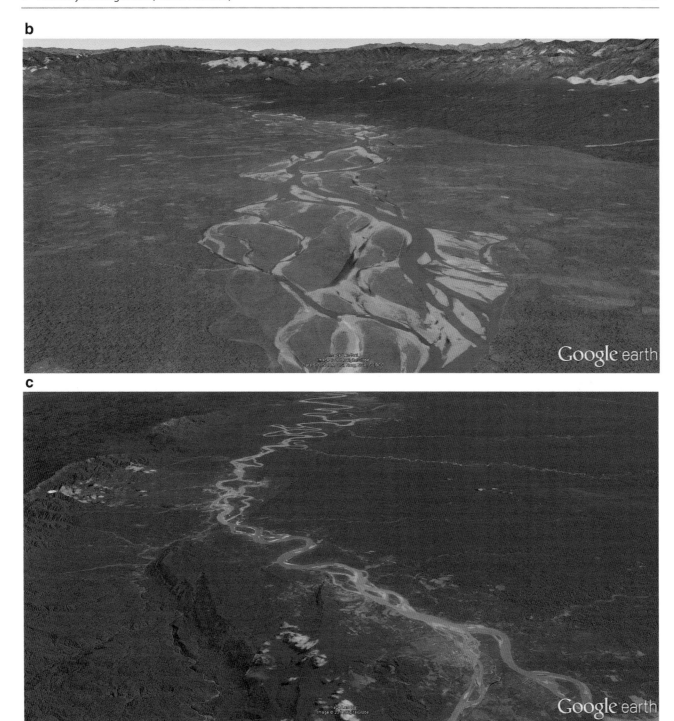

Fig. 9.30 (continued)

Fig. 9.30 (continued)

9 Forms by Flowing Water (Fluvial Features)

a

b

Fig. 9.31 (a) Back-cutting of ravines and small valleys into a higher terrace east of the Caspian Sea at *41°12′N* and *54°44′E*. The landscape shown is 5 km wide. (b) Farmland is under threat of intensive back-cutting of valley heads in the soft rocks of Andalusia, southern Spain, at *37°29′N* and *3°01′W*. Scene is about 6 km wide (Image credit: ©Google earth 2012)

a

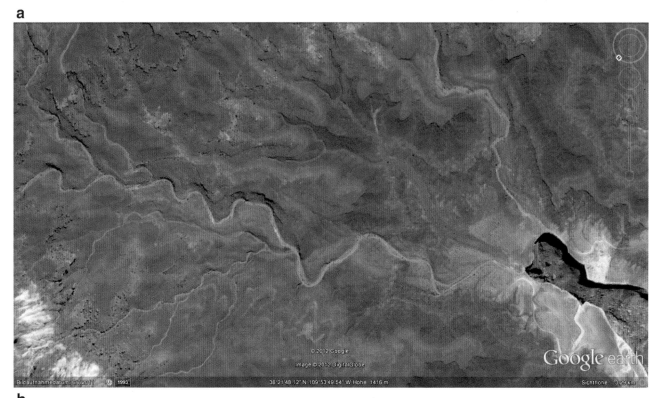

b

Fig. 9.32 (**a–c**) Canyon heads as tributaries of the Green River in Canyonlands National Park of Utah (USA), demonstrating the back-cutting which in the course of geologic time scales will erode the prominent white rim carbonate sandstone in the images (Image credit: (**a**) ©Google earth 2012, (**b**) S.M. May, (**c**) D. Kelletat)

9 Forms by Flowing Water (Fluvial Features)

c

Fig. 9.32 (continued)

Fig. 9.33 (**a**, **b**) The Canadian part of the Niagara Falls, called Horseshoe Falls. They show the back-cutting of this 55 m high and about 250 wide step. The result within the next several 10,000 years will be a transformation into river rapids further upstream (Image credit: ©Google earth 2012)

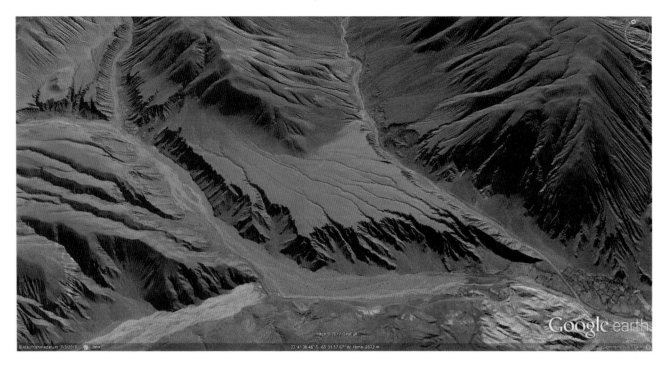

Fig. 9.34 Purmamarca valley in northwest Argentina (*23°41'S, 65°33'W*) and remains of a former valley fill. A period of sediment accumulation must have occurred before the river has incised into its own sediments (Image credit: ©Google earth 2012)

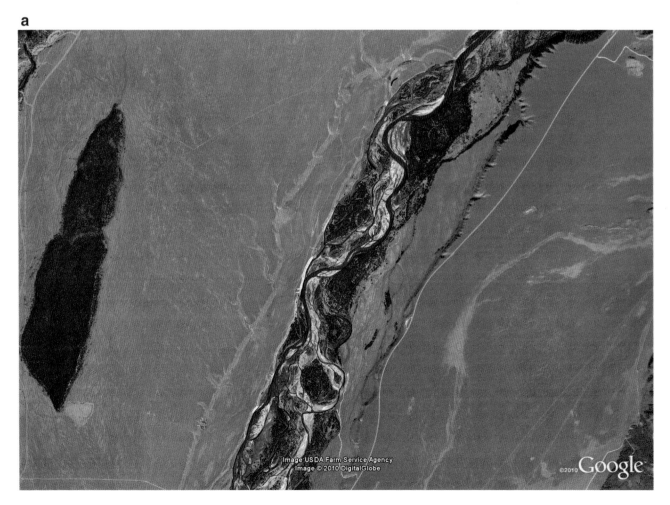

Fig. 9.35 (**a**) Wide terraces of the Snake River in Grand Teton National Park of Wyoming, USA, at *43°37'N* and *110°43'W*. Width of the scene of about 9 km. (**b**) Detail of (**a**). (**c**) Up to five terraces along a river bed in southern Patagonia (Argentina) at *42°21'S* and *70°32'W*. Scene is 21 km wide (Image credit: ©Google earth 2012)

234　　9　Forms by Flowing Water (Fluvial Features)

b

c

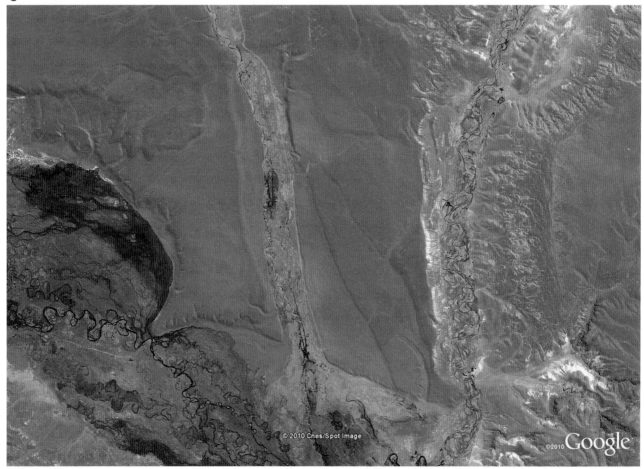

Fig. 9.35 (continued)

9 Forms by Flowing Water (Fluvial Features)

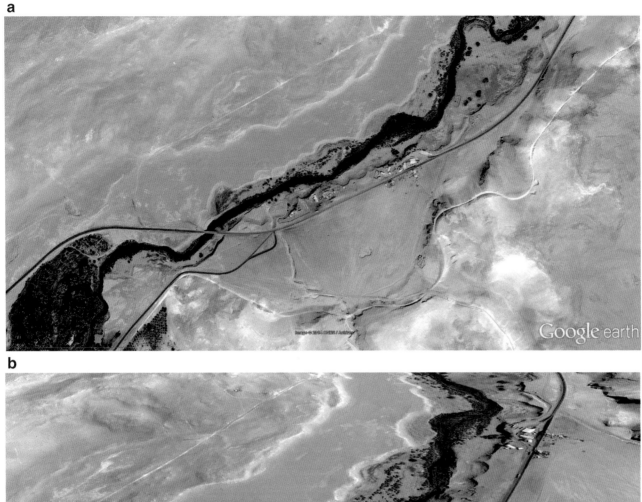

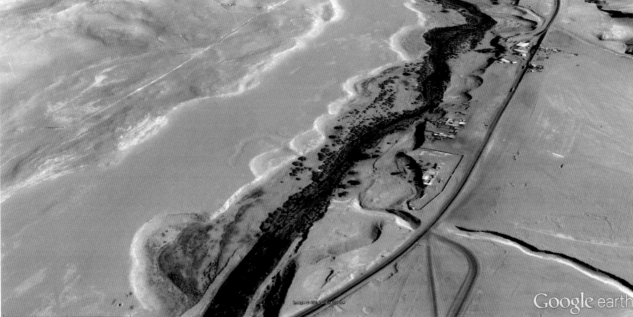

Fig. 9.36 Several terrace levels above the present floodplain along the Rio Loa, Chile (Image credit: ©Google earth 2012). (**a**) Section view to the west. (**b**) A tilted section with view to the northwest. (**c**) The same valley just a few kilometers further downstream; view is to the northwest

c

Fig. 9.36 (continued)

the foot of mountains, where the incline of the river bed suddenly changes into a remarkably lower angle. Here, the stream power immediately decreases and sediments are accumulated where the river valley enters the plain (i.e., the area of low relief). Debris fans mostly have the form of wide and smooth cones or parts of cones, and we usually see a bifurcation of the stream lines, ultimately resulting in the lateral extension of the fans (Fig. 9.37a, b). Debris fans are best developed in areas with strong physical weathering and where the transport of sediment load is not hindered by vegetation cover. Nearly all examples shown from satellite images are from arid landscapes around desert mountain belts or from very high mountains in the mid-latitudes with significant frost weathering (Fig. 9.38a–e).

Similar to debris fans, deltas are sediment bodies of fluvial deposition. Deltas are places where a river deposits its sediment load into a standing body of water, like a lake or the ocean, and loses its transport capacity. In contrast to most of the alluvial fans presented above, large deltas have an extremely low gradient of sloping. Deltas forming around river mouths in oceans represent the final sink for fluvial deposits and build the ultimate base level of rivers to which the longitudinal profile adjust. Here, littoral processes redistribute the fluvial sediments. Depending on the complex interplay between the river and its sediment load, the wave and tidal regimes, the dominating ocean currents as well as the bathymetric circumstances, different delta types are formed (Fig. 9.39a–d). However, all fluvial deltas found along the shorelines of the Earth's oceans are, geologically speaking, very young features because they can only have formed when the sea-level approached the present one in the mid-Holocene. The present sea level was reached between about 7,000 and 6,000 years ago as a consequence of deglaciation following the last glacial maximum ~18,000 year ago, when sea level dropped to around 120 m below the current level. However, some Holocene delta deposits (e.g., from the Mississippi or the Nile) have reached a remarkable thickness within this relatively short period of time. Since these sediments represent a rather large amount of weight on a limited area of the lithosphere, large deltas are also places of subsidence. In order to sustain delta progradation, the subsidence and auto-compaction of the sediment (removal of water and air from pore space) must be over-compensated by the new sediment brought by the rivers. Deltas can be found in all latitudes, but the largest are found in areas with high discharge, high precipitation and/or large amounts of sediment.

Fig. 9.37 (a) Single alluvial fans in the Death Valley National Park, California, USA. The large fan at the bottom of the image is nearly 5 km wide and is crossed by a road along a contour line. Typical bifurcation is seen as the river splits into several channels, causing deposition in all directions and, ultimately, forming the cone-like shape. (b) The western part of the Death Valley showing extensive alluvial fan systems extending from the surrounding mountain chains into the basin. Individual fans are grown together laterally. Again, typical bifurcation is present (Image credit: ©Google earth 2012). (c) Photo of a similar aspect as in (b) (Image credit: S.M. May)

Fig. 9.37 (continued)

Fig. 9.38 (a) A single alluvial fan about 400 m wide, deposited in a graben in Djibouti at *11°57′N* and *42°24′E*. (b) Alluvial fans in the Vacas valley, southeast of Aconcagua, Andes, western Argentina. The activity of the fans pushes the river to the opposite valley side. (c) In southern Mongolia, a little bit further to the east of (d), similar fans can be found near *45°10′N* and *96°24′E*. Image here is 42 km wide. (d) A colorful fan belt along a mountain range in northern China near *44°55′N* and *92°06′E*. Width of image is 35 km. (e) Each of these glacier-filled valleys deliver a wide alluvial fan at the foot of the mountains in central China. Scene is 24 km wide at *35°55′N* and *90°45′E* (Image credit: ©Google earth 2012)

Fig. 9.38 (continued)

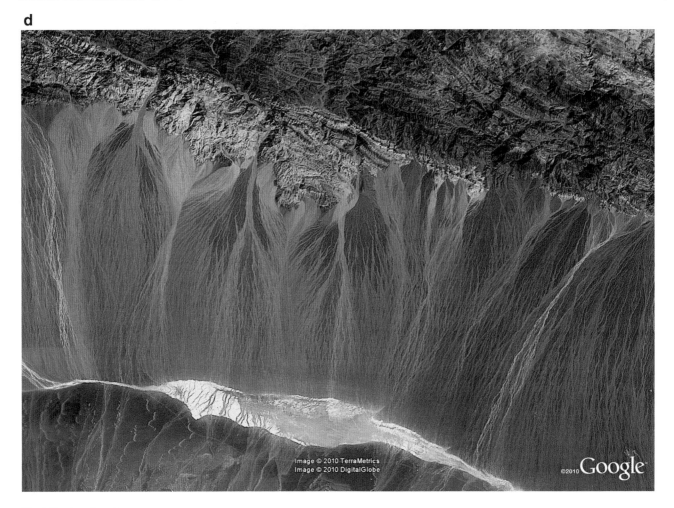

Fig. 9.38 (continued)

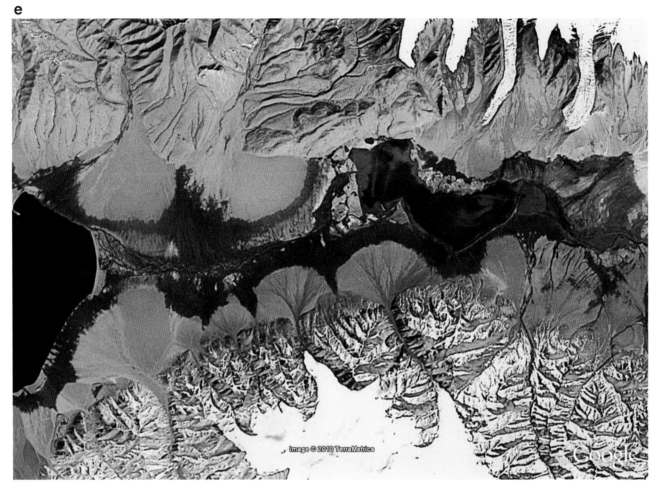

Fig. 9.38 (continued)

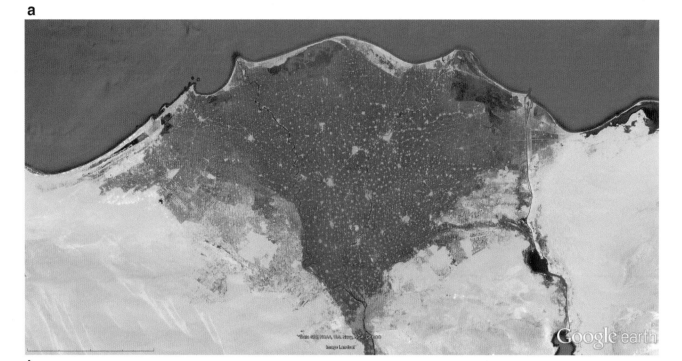

Fig. 9.39 (a) The Nile delta as a typical example for an arcuate delta, resembling the Greek letter Δ (eponymous for delta!) (b) Cuspate delta of Rio Grijalva at the north coast of Yucatán, Mexico, at around *18°33′N* and *92°39′W*. The scene is 23 km wide. The delta has been built by a series of beach ridges at both sides of the river mouth, where sediments have been distributed by longshore drift from the northern sector to both sides. (c) Rounded Yukon delta in northern Alaska (USA) with the center of the image at *62°52′N* and *164°05′W*. Width of the scene is 165 km. The dominance of the fluvial power can be explained by limited wave energy in often ice-covered arctic waters. (d) The classical form of a birdfoot delta is represented by the Mississippi River entrance into the Gulf of Mexico. Its main parts are built by natural levees along each of the river arms into the sea. The Holocene delta lies on top of a nearly 80 km wide older sediment fan formed by former deltas. (e) Along the east coast of New Zealand's South Island, blocked deltas are typical. Their seaward protrusion is hindered by strong waves and longshore currents. Scene at about *43°59′S* and *171°47′E* and is 110 km wide (Image credit: ©Google earth 2012)

9 Forms by Flowing Water (Fluvial Features)

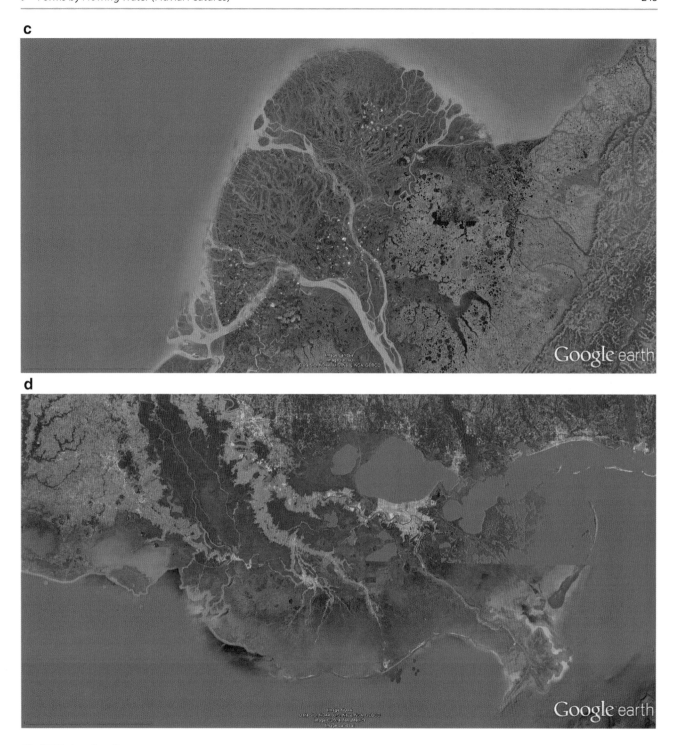

Fig. 9.39 (continued)

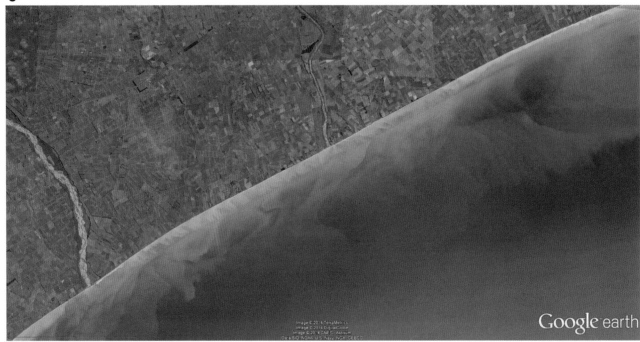

Fig. 9.39 (continued)

Further Readings

Bhattacharya JP, Giosan L (2003) Wave-influenced deltas: geomorphologic implications for facies reconstruction. Sedimentology 50:187–210

Brierley GJ, Fryirs KA (2005) Geomorphology and river management. Blackwell, Oxford

Correggiari A, Cattaneo A, Trincardi F (2005) Depositional patterns in the late-Holocene Po delta system. In: Giosan L, Bhattacharya J (eds) River deltas: concepts, models, examples, SEPM Special Publication 83. Tulsa, USA, pp 365–392

Dollar EJ (2000) Fluvial geomorphology. Prog Phys Geogr 24:385–406

Dollar EJ (2002) Fluvial geomorphology. Prog Phys Geogr 26:123–143

Dollar EJ (2004) Fluvial geomorphology. Prog Phys Geogr 28:405–450

Giosan L, Bhattacharya J (eds) (2005) River deltas: concepts, models, examples. SEPM Special Publication 83. Tulsa, Oklahoma (USA), 502 p

Graf GL (1988) Fluvial processes in drylands. Springer, Berlin/Heidelberg

Gregory KJ, Benito G (eds) (2003) Palaeohydrology: understanding global change. Wiley, Chichester

Gregory KJ, Macklin MG, Walling DE (2006) Past hydrological events related to understanding global change: an ICSU research project. Catena 66:2–13

Hardy RJ (2005) Fluvial geomorphology. Prog Phys Geogr 29:411–425

Jacobson R, O'Connor JE, Oguchi T (2003) Surficial geologic tools in fluvial geomorphology. In: Kondolf GM, Piegay H (eds) Tools in fluvial geomorphology. Wiley, Chichester, pp 25–57

Juracek KE (2014) Geomorphic changes caused by the 2011 flood at selected sites along the lower Missouri River and comparison to historical floods. U.S. Geological Survey Professional Paper 1798-H. Reston, Virginia (USA), 15 p

Knighton D (1998) Fluvial forms and processes – a new perspective. Wiley, New York

Kondolf GM, Piégay H (eds) (2003) Tools in fluvial geomorphology. Wiley, Chichester

Macklin MG, Benito G, Gregory HJ et al (2006) Past hydrological events reflected in the Holocene fluvial history of Europe. Catena 66:145–154

Meade RH (1996) River-sediment inputs to major deltas. In: Milliman JD, Haq BU (eds) Sea level rise and coastal subsidence. Kluwer, Dordrecht, pp 63–85

Newson MD (2006) 'Natural' rivers, 'hydrogeomorphological quality' and river restoration: a challenging new agenda for applied fluvial geomorphology. Earth Surf Process Landf 31:1606–1624

Newson MD, Pitlick J, Sear DA (2002) Running water: fluvial geomorphology and restoration. In: Perrow MR, Davy AJ (eds) Handbook of ecological restoration, vol 1. University Press, Cambridge, pp 133–152

Nezu I, Nakagawa H (1993) Turbulence in open channel flows. Balkema, Rotterdam

Oti MN, Postma G (1995) Geology of deltas. A.A. Balkema, Rotterdam

Piégay H, Grant G, Nakamura F, Trustrum N (2006) Braided river management: from assessment of river behaviour to improved sustainable development. In: Sambrook-Smith GH, Best JL, Bristow CS, Petts GE (eds) Braided rivers: process, deposits, ecology and management, International Association of Sedimentologists Special Publication 36. Blackwell Publishing, Malden, Massachusetts (USA), pp 257–275

Roberts HH (1997) Dynamic changes of the Holocene Mississippi River delta plain: the delta cycle. Journal of Coastal Research 13:605–627

Roy A, Lane S (2003) Putting the morphology back into fluvial geomorphology: the case of river meanders and tributary junctions. In: Trudgill S, Roy A (eds) Contemporary meanings in physical geography. From what to why? Arnold, London, pp 103–125

Thorndycraft VR, Benito G (2006) The Holocene fluvial chronology of Spain: evidence from a newly compiled radiocarbon database. Quat Sci Rev 25:223–234

Thorndycraft VR, Benito G, Gregory KJ (2008) Fluvial geomorphology: a perspective on current status and methods. Geomorphology 98:2–12

Planar Forms and Plain Forming Processes: Pediments/Glacis, and Peneplains (with Inselbergs)

10

> **Abstract**
> Planar forms as individual geomorpic forms may be very extensive, but are rather rare on Earth, because they usually require long geologic time spans to form (millions or tens of millions of years). Nevertheless, there are typical planar forms according to climate provinces like the glacis or pediments in arid and semi-arid environments along the feet of mountains or mountain chains. The largest planar forms and the oldest ones are the peneplains (sometimes with inselbergs as remnants of a former more differentiated relief type). Their extension may reach several hundred kilometers across as in eastern Africa, but they also have been formed in other latitudes in former climates similar to the warm and semi-humid one of East Africa. Many of these are in a state of incision and destruction by rivers due to climate change and tectonic uplift, tilting and faulting.

As a strong contrast to valley incision and linear erosion by flowing water, plains occur in different shapes and climates, mostly also formed by the influence of flowing water. Some plains are just the extension of debris fans stretching far into plains, and the debris becomes smaller and smaller over a longer transport distance (Fig. 10.1).

In arid landscapes, many mountain belts and sometimes also insular hills are surrounded by smooth slopes away from the mountains, and these slopes are free of any other morphologic feature. They are the result of the back-cutting of mountain ranges and are mostly composed of hard rock under a sheet of rock debris; in most cases, the debris consists of unsorted, partly rounded to angular material, also containing some large boulders. These deposits are called "fanglomerates" and result from extreme but rare flash floods which transport the rock debris out of the mountains without following particular linear pathways of transport. Intense erosion and abrasion is affecting the hard rock below. This down- and back-wearing produces a fringe of "pediments" (areas of transportation and erosion, and predominantly without sediment cover) and more distal "glacis" (areas of predominant deposition and sediment cover) around mountains in arid and semi-arid landscapes, and eventually all the mountains will be cut back and disappear completely. Where two sloping pediments, expanding from both sides of a mountain range by ongoing pedimentation, meet with their upper parts at a drainage divide, a "pediment pass" is created. Very good examples of pedimentation can be seen in the arid areas of Nevada and eastern California as well as in Arizona and New Mexico of USA, but also along the northern belt of the Sahara Desert in Africa or at the foot of mountain chains in Asia, Australia and South America (Figs. 10.2, 10.3 and 10.4b). If long-term climatic conditions change (e.g. precipitation and runoff diminishes), the debris cover may become too thick to be moved, and the process of pediment formation stops.

Besides pediments and sediment-filled depressions, so-called "peneplains" represent a different form of plains. These are extremely wide plains that extend across a number of different rocks and rock structures. In most cases, they are covered by a thick layer of weathered material and deep soils. Due to their long unchanging forms over the geological time scale, their genesis has been under debate for more than 100 years. There is a loose agreement that the process of their formation involved a long time of chemical weathering activities under semi-humid to humid climates which have

Fig. 10.1 An outwash plain around the southern Sinai mountains in Egypt (Image credit: D. Kelletat)

affected all rock types over millions of years. The perfect plain form is only preserved where tectonic movements are (almost) absent and have not lifted or tilted the landscape. Residual outcrops as *kopjes* (a term for small rocky residual hills in South Africa), or *inselbergs*, can still be found in peneplains or planation surfaces (Figs. 10.5a, 10.6 and 10.7).

Fig. 10.2 A 20 km wide pediment north of Las Vegas, Nevada, USA, seen from an airplane. (Image credit: D. Kelletat)

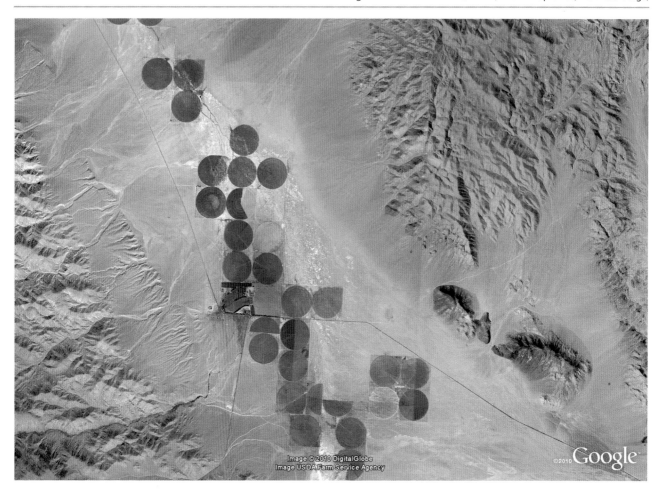

Fig. 10.3 Pediments in southwest Nevada, USA, from both mountain belts at about *37°30′N* and *117°54′W*. The left one has been disturbed by a north–south trending fault line, now forming a dissected step in the pediment. As a scale, the round irrigated areas have diameters of 800 m (Image credit: ©Google earth 2012)

a

Fig. 10.4 (**a**) Pediments in eastern Afghanistan at *32°38′N* and *61°25′E*, in a 32 km wide scene. (**b**) Detail of the same landscape shows the continuous diminishment of mountains between the pediments in the surrounding areas (Image credit: ©Google earth 2012)

b

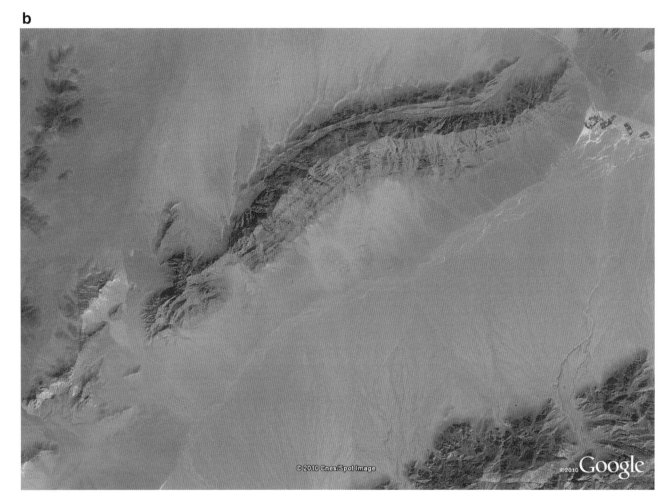

Fig. 10.4 (continued)

Fig. 10.5 (**a**) This image shows a peneplain in eastern Africa in the Tsavo National Park near *2°29'S* and *39°01'E*. This is a 50 km wide section, flat from horizon to horizon. Even the waterways from annual monsoonal rain could not carve into this plain because there is no inclination. Some residual mountains can be seen in the far distance (Image credit: D. Kelletat). (**b**) In this tilted aspect from eastern Africa, we can imagine how the remnants of a former relief are slowly disappearing in the wide peneplain (Image credit: ©Google earth 2012)

Fig. 10.6 In the wide peneplains of central Australia, Uluru (Ayers Rock) stands nearly 400 m of relative height above the surface and appears as an insular feature in the landscape. (**a**) Image taken from about 10 km distance, view is from the north (Image credit: D. Kelletat). (**b**) This image is taken from a distance of about 30 km, looking south (Image credit: ©Google earth 2012)

Fig. 10.7 (a) View from the inclined rocky pediment of Kata Tjuta (the Olgas), 35 km west of Uluru. In the background, the wide horizontal peneplain of central Australia can be seen. (Image credit: D. Kelletat). (b) Western part of Kata Tjuta with Uluru in the background. The pediments are the peripheral and lesser inclined sections around the beehive-formed rock in the image foreground (Image credit: ©Google earth 2012)

Further Readings

Atkinson D (2004) Weathering, slopes and landforms. Hodder Arnold, London

Ehlers EG, Blatt H (1982) Petrology, igneous, sedimentary, and metamorphic. W.H. Freeman & Co., New York

Goudie AS, Viles H (2008) Weathering processes and forms. In: Burt TP, Horley RJ, Runsden D, Cox NJ, Goudie AS (eds) Quaternary and recent processes and forms. Geological Society, London

Monroe JS, Wicander R (2000) Physical geology: exploring the Earth, 2nd edn. West Publishing, Minneapolis/St. Paul

Plummer CC, McGeary D, Carlson DH (2003) Physical geology, 9th edn. McGraw-Hill, Dubuque

Press F, Sievers R, Grotzinger J, Jordan T (2004) Understanding earth, 4th edn. W. Freeman & Co, New York

Forms by Wind (or: Aeolian Processes): Deflation and Dunes

11

Abstract

Compared with the erosive force of moving glaciers and flowing rivers, wind seems to be a less effective agent when it comes to the evolution of landscapes and the generation of landforms. While this idea may be true for the destructive aeolian process of deflation and its resulting landforms, the depositional aeolian processes such as dune formation are responsible for a remarkable variety of important (large and small) landforms that characterize wide regions on Earth. Parabolic coastal dunes, migrating barchan dunes and sand mountains such as the enormous star dunes are just a few examples that can be easily identified with Google Earth. Erosive aeolian landforms such as yardangs also characterize large parts of deserts and they are shaped by the corrasive power of saltating sand grains over long periods of time.

While wind is present everywhere on Earth, landscapes which are dominated by aeolian landforms or deposits are not. Whether wind can sculpture the Earth's surface depends on its ability to erode and transport material, the availability of sediment to be transported and, ultimately, the depositional environment. Even in very windy areas, a dense vegetation cover (e.g., in humid climates with moderate temperatures) can hinder the wind from affecting the landscapes. Regions with dominant aeolian landforms are thus mainly found in arid or semi-arid climates, and of course deserts are areas of typical aeolian landforms. The largest deserts on Earth can be found in Africa (the Sahara, Namib, Kalahari and Danakil), on the Arabian peninsula (the Rub al-Chali or the "Empty Quarter", and Nefud), in Iran (the Lut and Dasht e Kevir), in central Asia (the Takla Makan, Gobi, and areas around Lake Aral, and Thar in India and Pakistan), in South America (Atacama), in North America (the Sonora, Chihuahua, Mojave and Great Basin), and Australia (the Great Victoria, Great Sandy, Gibson and Simpson).

Sand deserts are usually termed "*erg*", and sand dunes dominate these desert areas. However, deserts are not a synonym for dune fields, and the presence of dunes is not the definition of deserts. Moreover, desert landscapes are not restricted to hot climates – they are also be present in very cold (and in many cases also arid) Arctic and Antarctic climates.

Less than 50 % of the area of most deserts mentioned above consist of dune fields; large parts of these deserts are dominated by landscapes of bare rocks, stone pavements, and rock debris of all sizes (e.g., "*reg*" in Arabic, "*dasht*" in Persian, or "*gobi*" in Chinese). In reg deserts, the term "hamada" refers to a desert with bare rocks and coarse rock debris, and "*serir*" typically describes deserts with smaller rock fragments like pebbles. Both terms derive from the Arabic and are used for desert areas of the Sahara.

Similar to other forming agents, we may differentiate the aeolian landforms into destructive forms and depositional forms. Wind does not only accumulate sediment in the form of dunes (mainly sand) or loess (mainly silt), it also erodes and transports fine sediment grains from open surfaces with sparse vegetation. Deflation is an aeolian erosional process, where fine sediment is simply blown away. Deflation may result in the degradation of intensely used landscapes (Fig. 11.1); in the formation of bowl-shaped blowout depressions, or the lowering and enlargement of existing ones (Fig. 11.2a–c); as well as in the evolution of stone pavements (Fig. 11.3, see also Fig. 11.5a). In general, the coarsest sediment grains transported by wind are pebbles of a few centimeters in diameter, though only moved by rolling or saltation during extreme storm conditions. Where only the coarser sediment grains such as pebbles remain at the surface

Fig. 11.1 Where vegetation is destroyed due to overgrazing by sheep in Iceland, wind can deflate fine sediment grains, resulting in the erosion and desertification of landscapes (Image credit: D. Kelletat)

and the finer sediment fractions are blown out, stone pavements develop, which in turn, protect the underlying finer material from erosion (Fig. 11.3).

Saltation is the most common transport mode for sand grains. Once set in motion by wind, the sand grains saltate over the ground and impact other grains, which becomes incorporated into the saltation process. Angular sand grains produced by mechanical/physical weathering are rounded by sediment transport such as the steady movement by waves at a beach, by flowing water, or by wind transport in the form of saltation. Since only the most resistant minerals survive a longer period of transport, a selection of more resistant minerals occurs with time. Therefore, most of the sand in deserts and at beaches in the world (up to 95 %) consists of quartz grains (silicon dioxide, SiO_2). However, in tropical and subtropical latitudes, where coral reefs are abundant along coastlines, beaches often consist of calcareous sand-sized fragments. In addition to the common transport mode of saltating sand grains, wind also induces suspended sediment transport, although finer sediments such as silt and clay may be more cohesive and more difficult to be set in motion.

Where large amounts of fine sediment such as silt are transported in suspension, dust storms may form (Fig. 11.4). In most cases, the term "sand storm" is misleading since the typical sediment-loaded clouds of these are dominated by the finer sediment, silt.

Where sediment particles such as sand grains are transported with the wind, aeolian erosion acts in the form of corrasion. Corrasion occurs where wind-blown particles collide with hard rock and thereby polish and sculpt it into different forms. Corrasion may abrade rocky obstacles into "ventifacts" (i.e., new forms made by wind) and even "mushroom rocks" (Fig. 11.5a–c). It is most effective within the first meter above the surface, particularly in the section without ground friction. Here, saltating sand grains affect the rock surfaces and successively abrade lower rock sections. If the wind steadily blows from one direction over a long period of time such as in the Earth's trade wind belts, the landscape may be decorated by sculptures of the same pattern with similar dimensions. In areas of drifting sand, rock outcrops are sculpted along their weakest parts which normally are fissures and bedding planes. Elongated forms and whaleback forms are most abundant and in areas where less resistant sedimentary rocks are eroded, the forms are called "yardangs". These forms are well known, for instance, from the Iranian and Chinese deserts (Figs. 11.6a–c and 11.7a–c).

The most important and best-known landforms made by wind are those generated by the deposition of mostly fine to medium sand, i.e., dunes. Dunes exhibit many different forms, depending on wind velocity, wind direction (and the permanence of wind directions), sand availability, surface roughness and vegetation density, and sometimes man-made disturbances. Single dunes may form within a short period of time and the shape of single dunes may change within hours (e.g., during stormy winds). However, landscapes of large dune systems and sand seas need rather constant conditions over longer time periods to form. While wide dune areas such as sand seas are typical for sand deserts, narrow and rather isolated (but sometimes very long or large) dunes can be found in all latitudes, such as along beaches worldwide. Here, constant sediment (sand) availability and onshore blowing winds prevailing throughout the year may generate coastal dunes, which are more independent of climate and vegetation density. The size or positive nourishment of these dunes depends on the extension of the dry parts of the beach, which also depends on the tidal range. During low tides, exposed parts of the beach may dry out. Winds can move the sand grains inland, where they accumulate and form dunes as permanent or semi-permanent deposits as wind drift decreases or friction increases. Similarly, dunes may also be found independent from coastlines bordering sand-dominated rivers. Beaches are also best places to study initial dune forms, small ripple fields as well as the process of saltation of sand grains by wind.

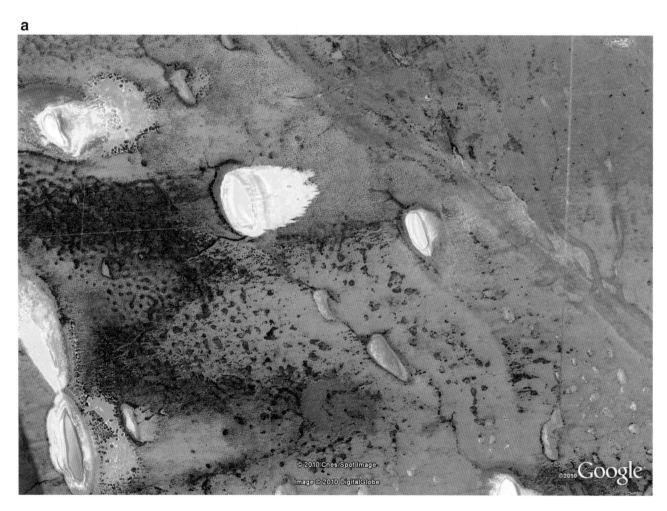

Fig. 11.2 (a) Blowout depressions showing deflation in Patagonia (Argentina) at *51°17′S* and *70°30′W*. Wind direction is from the west and is illustrated by the blowout of salt. Section is about 5.2 km wide. (b) Long tails of salt show the direction of strong winds, forming blowout depressions in southern cold and dry Patagonia, Argentina (*51°52′S* and *69°49′W*). Scene is about 20 km wide. (c) Deflation depressions between old dunes in the Shark Bay area of Western Australia at *25°47′S* and *113°40′E*. Scene is about 5 km wide (Image credit: ©Google earth 2012)

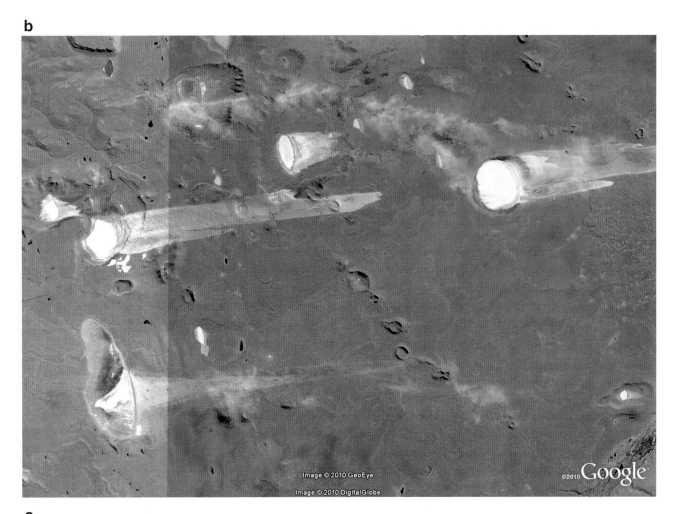
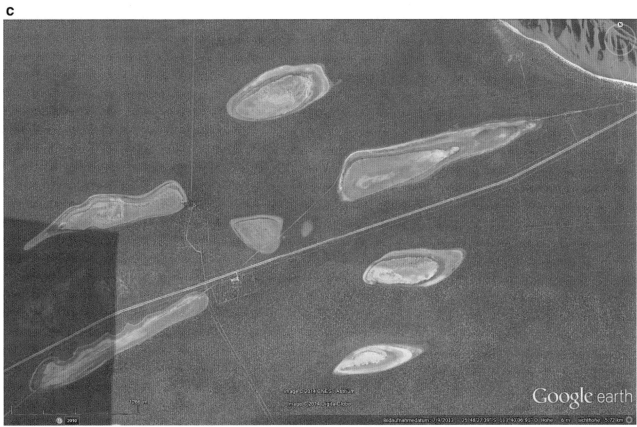

Fig. 11.2 (continued)

Fig. 11.3 If all fine grains are blown away by wind during the process of deflation, only the larger fragments are left and concentrated on the surface, forming a stone pavement. Due to its roughness, it protects the underlying fine sediments from deflation. The picture shows a nice example of a desert pavement (with desert varnish on each debris) in the Negev Desert of Israel (Image credit: D. Kelletat)

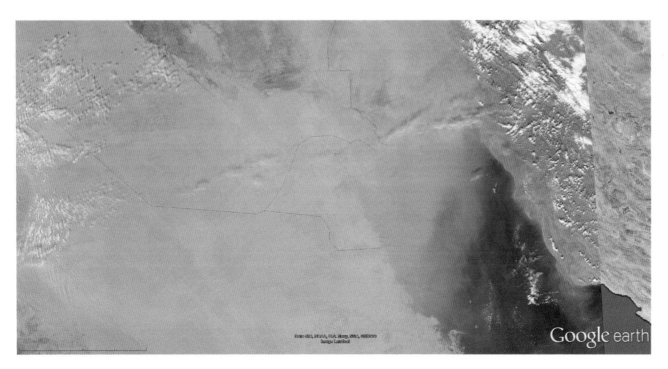

Fig. 11.4 Dust storm over the Middle East (Saudi Arabia, Kuwait, Iraq and Iran), captured by MODIS on NASA's Aqua satellite (Earth Observatory NASA) early January 2013. Center of image is at *29°23′54.41′N* and *47°42′34.26′E*; image is 1,000 km wide. Large sand seas as well as alluvial sediments brought by the Euphrates and Tigris Rivers are sources of fine sediments for frequently occurring dust storms in this desert region. See http://earthobservatory.nasa.gov/NaturalHazards/view.php?id=80112 for Google Earth file

Fig. 11.5 (a) Wind may sculpture and polish ventifacts in the zone above strongest ground friction (wind direction is mainly from the left). This example is from meltwater debris in southern Iceland, which is barren of vegetation in this windy region. The debris is predominantly composed of hard basalt. (b) Wind transporting sand grains close to the surface polish rock outcrops and sculpt rock surfaces.

Fig. 11.5 (continued) This rock of less than 1 m in diameter at the west coast of Portugal is made of limestone, where wind sculpting is ubiquitous in the coastal landscape. Quartz grains are readily available from beaches as corrasive tools. (**c**) Perfectly formed mushroom rock about 1.5 m high in Death Valley National Park, California, USA. The shape of the rock illustrates how and where the corrasive force of wind is most effective. While the base of the rock is polished but still wide, corrasion is most effective above the zone of strongest ground friction, where sand grains are transported with maximum speed. The rock is transformed into a thin column in its middle part. Just a few centimeters higher, the angular shape of the rock documents the upper limit of corrasion as sand grains are not saltating that high (Image credit: D. Kelletat)

The following series of images document the large catalogue of aeolian deposition and dunes and dune landscapes of the world, starting with small forms of sandy ripple marks (Fig. 11.8a, b). Different dune forms have very specific shapes, generated due to different wind- and surface-related circumstances. The elongated arms of parabolic dunes are usually anchored by vegetation and point windward, forming an U-shaped dune structure. Parabolic dunes typically develop from blowouts along coastlines, or may result from the coalescence of at least two sand shadows leeward of obstacles (typically single plants). Parabolic dunes are typical for many coastlines of the world, even in temperate and cool climates of the mid-latitudes, but also occur in semi-arid landscapes (Fig. 11.9). Minimal vegetation cover is a prerequisite for the generation of parabolic dune forms.

The classical migrating dune form found in relatively flat landscapes with sparse vegetation cover and with a lot of sand available is the "barchans" dune (crescent dune, crescent-shaped dune, horseshoe dune). The process of barchan dune formation and migration is easy to understand. A bare sand hill (irregular or symmetrical) has less sand in its outer section, and most of the sand is concentrated in its center. Therefore, during the same time period and by the same wind velocity (with the same transport capacity), the outer dune sections will move faster than the main body of sand. A migrating semi-circled and crescent-like dune form develops,

a

b

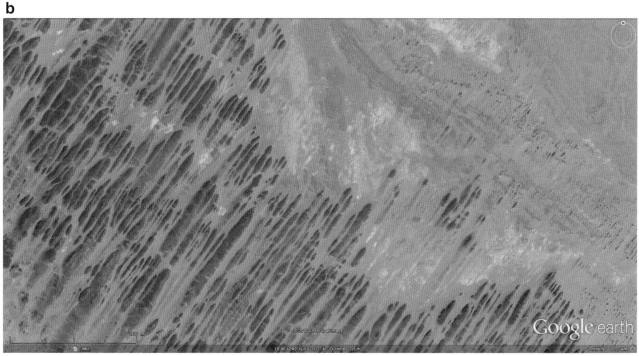

Fig. 11.6 (a) Long sand tails of migrating barchan dunes (image center), wind-eroded (corrasion) cuestas and joint patterns in Chad at about *19°14′N* and *19°22′E*. Scene is about 130 km wide. (**b**, **c**) Whaleback rocky outcrops formed by sand corrasion along a main joint pattern in Chad (**b**, at *18°46′N, 19°31′E*) and Niger (**c**, at *22°34′N, 11°4′E*) (Image credit: ©Google earth 2012)

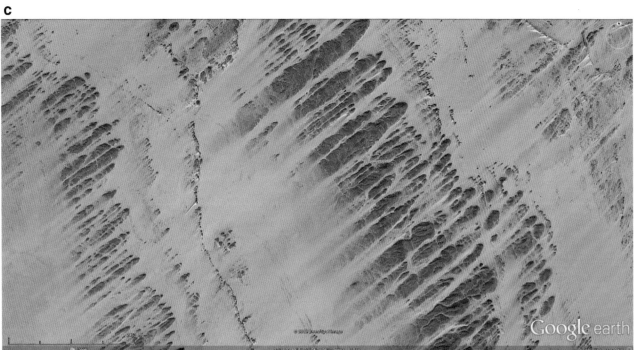

Fig. 11.6 (continued)

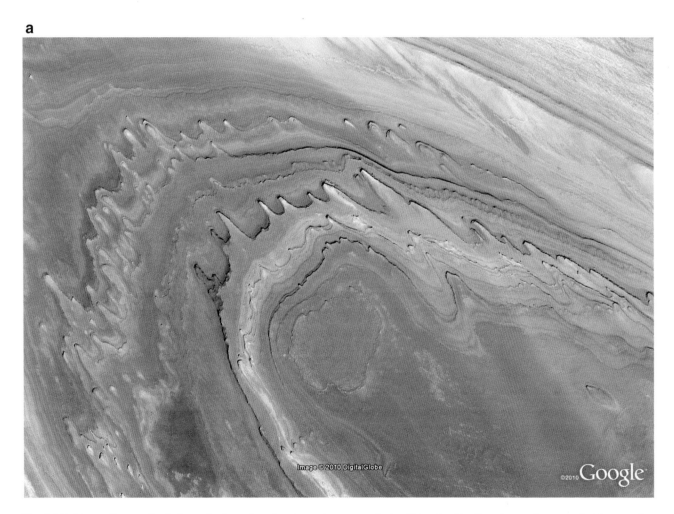

Fig. 11.7 Linear forms of wind erosion/corrasion (yardangs in "*statu nascendi*") (**a**) and perfectly formed yardangs (**b, c**) in China at 38°09′N, 91°47′E, 38°14′N, 93°24′E and 37°58′N, 93°59′E, respectively. Old and soft sedimentary rocks are eroded by corrasion with dominant wind direction from the north-northwest (Image credit: ©Google earth 2012)

b

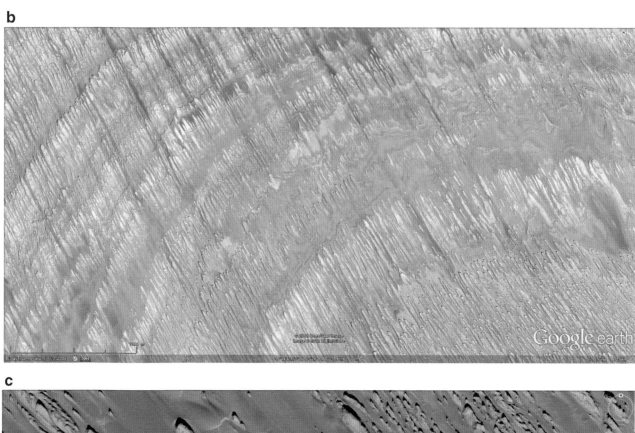

c

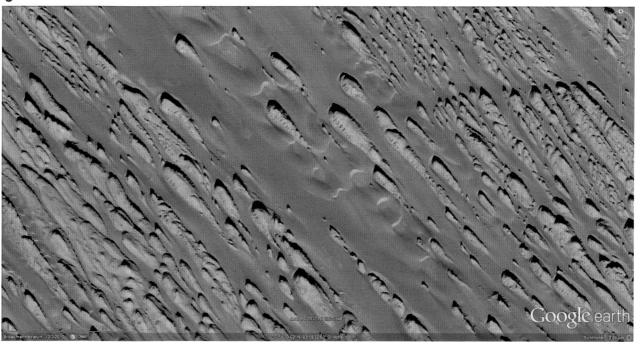

Fig. 11.7 (continued)

Fig. 11.8 (a) Small 1–2 cm high ripple marks (several centimeters to decimeters wide) as typical secondary (and short-lived) depositional features on dune surfaces. This one is from Fuerteventura Island, Canary Islands, Spain (Image credit: D. Kelletat). (b) Typical sand avalanches at the steep leeward (*downwind*) slope of a dune. This process is of great importance for the migration of dunes. Sand grains are shifted/transported to the top of the dune by saltation, where they accumulate behind the crest due to decreasing wind speeds. Ongoing sand accumulation produces instabilities along the dune crest, and gravity forces the sand to slide downhill. (c) Sand ripples on top of the flat surface of Lake Frome, which is characterized by salt efflorescence (central Australia). Ripple marks are just a few centimeters high, but resemble forms of larger dunes (Image credit: S.M. May)

Fig. 11.8 (continued)

11 Forms by Wind (or: Aeolian Processes): Deflation and Dunes

c

Fig. 11.8 (continued)

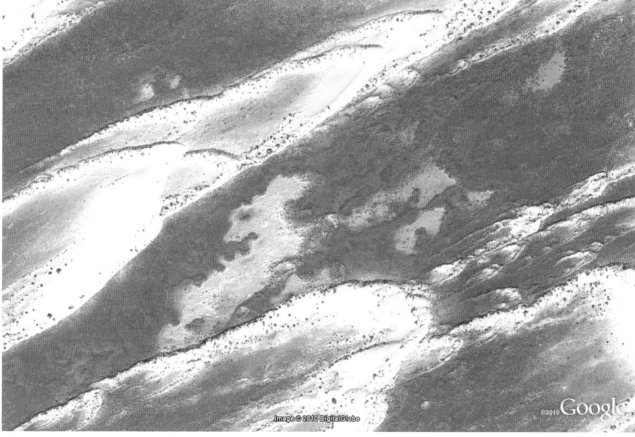

Fig. 11.9 (a) Parabolic dunes in a semi-arid and sparsely vegetated landscape of New Mexico, USA, at the eastern margin of the White Sands National Monument dune field at *32°54′N* and *106°12′W*, migrating from the left to right. Image is about 4.3 km wide. Parabolic dunes point with their convex nose (dune front) downwind, which is in contrast to barchan dunes. The elongated arms of parabolic dunes are anchored by sparse vegetation. (**b**) Detail of parabolic dunes from the same area in New Mexico, at *32°45′N* and *106°15′W*. Image is ~1 km wide. (**c**) Parabolic coastal dunes at the coast of Santa Catarina Province, southeast Brazil (*30°26′S* and *50°20′W*). Wind direction is from upper right (northeast), and scene is 1.5 km wide. More inland, a field of barchans or barchanoids is formed by the same wind direction. Typically, parabolic dunes are associated to blowout depressions along coastlines or in semi-arid landscapes; however, they may also form where plants trap sand particles on their leeward side, and neighboring sand shadows grow together (Image credit: ©Google earth 2012)

c

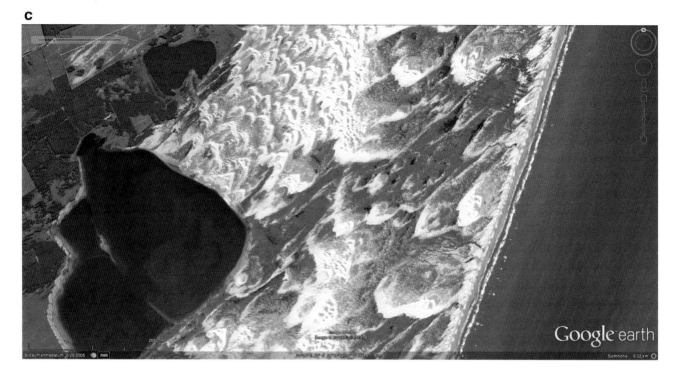

Fig. 11.9 (continued)

with two outer sections (sand tails or branches) travelling ahead and downwind of the main central dune body (Figs. 11.10a–d and 11.11). Barchan dunes often appear in larger groups and may be linked together to form "barchanoid" dunes (Fig. 11.12a–f). Particularly in areas with pronounced wet and dry seasons, the successive movement of barchan or barchanoid dunes may be preserved in the form of striation or striped patterns. In the wet season, the groundwater level is high, and the sand in the lower part of the dune is too wet to be transported; plants may anchor the windward contour of the dune. While the main body of the dune migrates during the dry season, the anchored sand area is left behind and the contour of the former windward dune slope is preserved (Figs. 11.12b and 11.13). In contrast to the migrating barchan dunes, the lee-dunes are bound to stationary obstacles such as rocks, plants or mountain chains. If wind blows from one dominant direction, stationary lee-dunes may form on the leeward side of these obstacles (Fig. 11.14).

Under special wind conditions (i.e., relatively constant wind directions over longer time periods), dunes may become organized into rather linear features as well, either as longitudinal (roughly parallel to the dominant wind direction; "*seif*" dunes) or transverse (perpendicular to the dominant wind direction) dunes. The following series of satellite images illustrate different types and forms of linear dunes, and some of them belong to dune systems more than 100 km wide. Examples are from different parts of Africa, Arabia and Australia (Figs. 11.15a–e and 11.16a–c). Barchanoid dune fields may be transformed into linear dunes and transitional patterns with all kinds of intermediate forms can be found (Fig. 11.17a, b). Linear dune forms may reach lengths of 100 km, and the fields of linear dunes in the Sahara and in the Australia deserts may cover tens of thousands of square kilometers. Many longitudinal dunes such as in the outback of Australia are very old dune forms.

Finally, "star dunes" can be found only in a few regions on Earth (e.g., Namibia in southwest Africa, southern Peru, or the "Empty Quarter" in southern Saudi Arabia and southwest Oman). Star dunes may be described as sand mountains (Fig. 11.18) and can be hundreds of meters high and many kilometers wide, with very different forms. The dune crests of star dunes usually stretch into all directions, and this typical star-like appearance is eponymous for the term star dunes (Fig. 11.19a–e). Changing wind directions without a particularly dominant wind direction and an extreme amount of available sand are prerequisites for the formation of star dunes.

In large dune fields, constant sand movement takes place and billions of tons of sand per year are transported. However, the geomorphological pattern may stay very constant for decades or centuries, and the dunes may not change in number, size or general shape. To maintain these conditions (i.e., an equilibrium between sand input and sand loss in a dune system), continued input of sand into the system is

Fig. 11.10 (a) A single barchan dune in Namibia, migrating from south to north (*26°54′33.43′S* and *15°20′57.49′E*). Width (east–west) and length (north–south) of the barchan dune is ~0.5 km. (b) Barchan dunes migrating downwind from south to north in Namibia at *26°56′S* and *15°20′E* (c) Along the eastern fringe of a wide coastal dune belt in northern Namibia, between *19°28′S, 12°53′E* and *19°40′S, 13°00′E*, barchan dunes are growing together to form garlands of dunes (i.e., barchanoid dune forms). Image has a width of about 4 km, with wind from the south. (d) A field of single barchan dunes in the Atacama Desert in Peru at about *14°00′S* and *75°53′W*. Image is 2 km wide. Wind direction is from upper left side (Image credit: ©Google earth 2012)

c

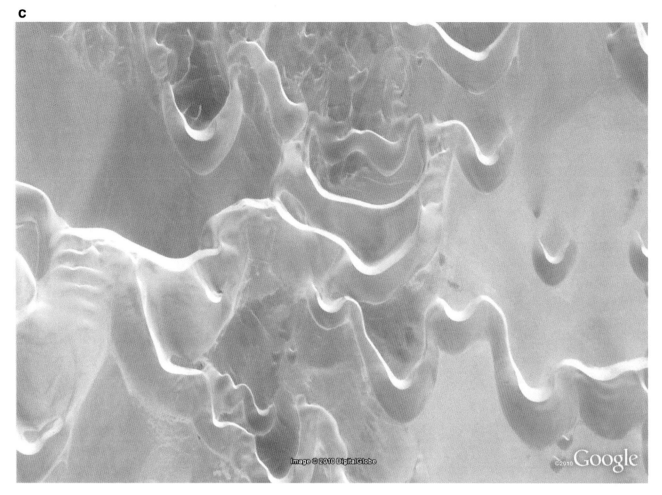

Fig. 11.10 (continued)

Fig. 11.10 (continued)

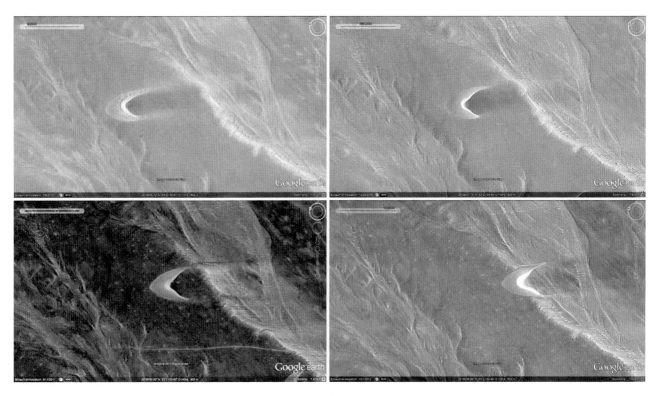

Fig. 11.11 A single and migrating barchan dune in western Pakistan (*29°00′N* and *63°21′E*), captured at four successive points in time (8/3/2010, 24/12/2010, 11/9/2011, 7/10/2012). Direction of movement is from northwest to southeast. Width of dune center is ~50 m. Additional further migrating barchan dunes are present in the direct vicinity, originating from a larger dune field in the northwest (Image credit: ©Google earth 2012)

Fig. 11.12 (a) East of the Amazon River mouth in northeast Brazil, fields of barchan or barchanoid dunes stretching from the coastline in an area of high precipitation and with considerable vegetation cover. In this case, longshore drift delivers sand from the Amazon, and sand availability is very high; together with constant trade winds from east-northeast this results in desert-comparable conditions and the dunes are shifting from the coast (upper right, east-northeast) inland. Site is at 2°33′S and 42°56′W, and scene about 8 km wide. (b) In this detail of (a) we see evidence of a successive movement of the barchanoid dunes in the form of a striation or striped pattern. In the wet season, groundwater level is high and the sand in the lower part of the dune is too wet to be transported and plants may anchor the local contour of the dune. The main body of the dune migrates during the dry season, and the anchored sand area is left behind preserving the contour of the former windward dune slope. (c) In this scene from the same area (about 1 km wide) the asymmetrical shape of the barchanoids with gently inclined windward (right) and steep leeward (left) slopes clearly can be identified. (d) Barchanoid dunes in White Sands National Monument, New Mexico (USA), at ~32°53′N and 106°15′W. The dunes are mainly composed of gypsum crystals, formed by evaporation and deflated from ephemeral lakes (i.e., Lake Lucero, a salt pan/playa) in the Tularosa Basin. The location is just a few kilometers to the west of Fig. 11.9a, where the barchanoid dunes are transformed into parabolic dunes due to semi-arid vegetation and reduced sand availability. (e, f) Barchan and/or barchanoid dunes in southeast Saudi Arabia at 22°39′N and 54°22′E. (e) shows a detail of 6 km width with giant barchan dunes, some nearly 150 m high and covered by smaller barchan or barchanoid dune forms (Image credit: ©Google earth 2012)

274 11 Forms by Wind (or: Aeolian Processes): Deflation and Dunes

b

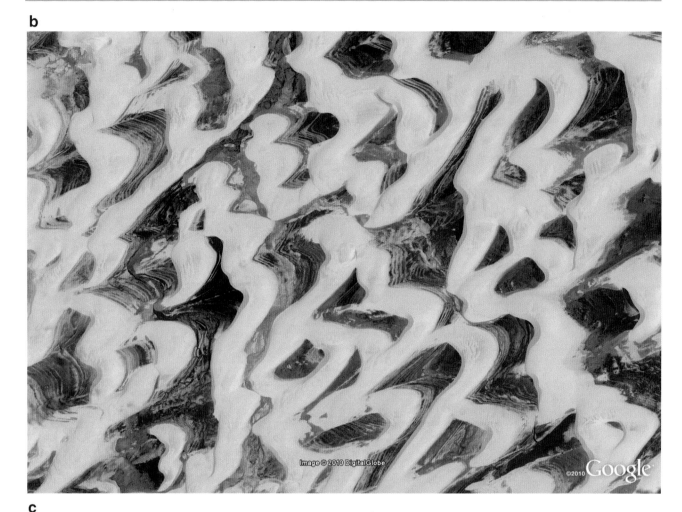

c

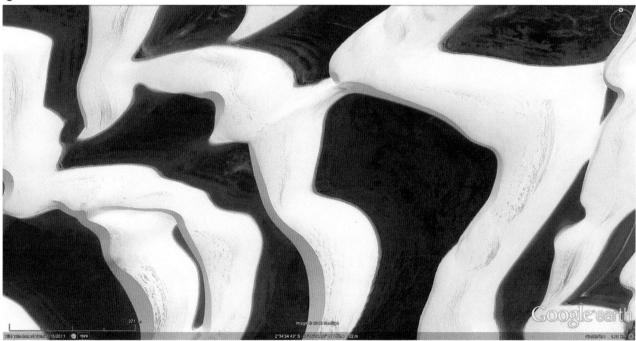

Fig. 11.12 (continued)

d

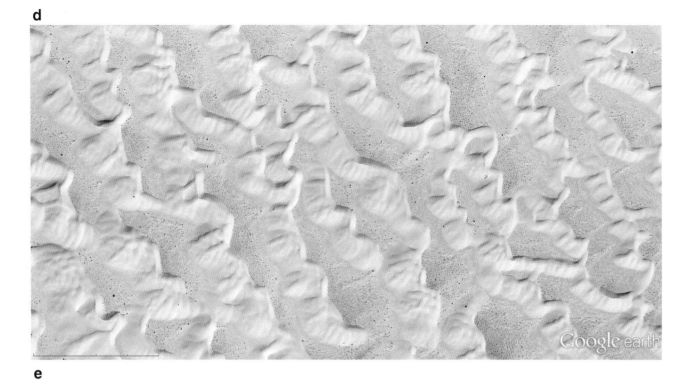

e

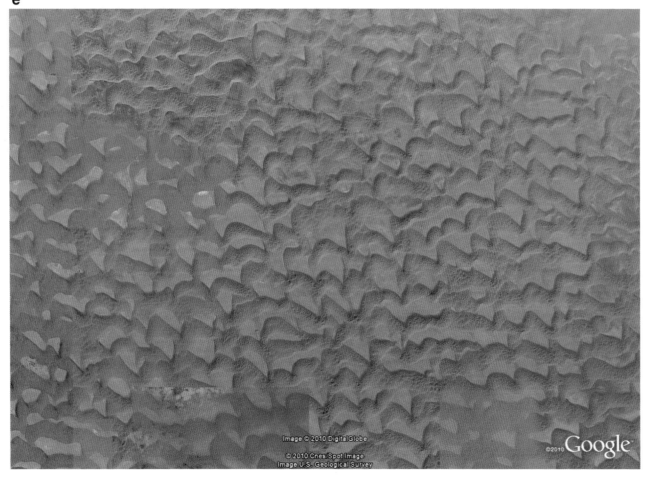

Fig. 11.12 (continued)

f

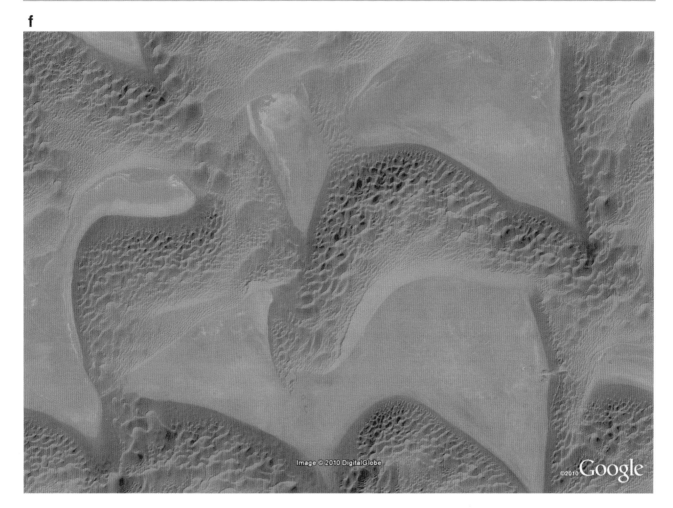

Fig. 11.12 (continued)

Fig. 11.13 Sand striations or stripes document the successive migration of a single barchan in northeast Brazil, south of the Amazon River, at *2°44′S* and *42°23′W*. The stripes may be formed due to seasonal changes in groundwater levels (wet and dry season) and the related changing amount of sand trapped along the windward and convex margin of the dune body. The scene is about 2.5 km wide (Image credit: ©Google earth 2012)

Fig. 11.14 Lee-dunes bound to rock outcrops in northern Niger, at *22°23′N* and *10°48′E*. Scene is 10 km wide (Image credit: ©Google earth 2012)

Fig. 11.15 (a) Parallel linear dunes formed by winds from west-northwest in central China at ~*36°59′N* and *93°59′E*. Width of image is ~25 km. (**b**, **c**) The northeast trade winds result in the extensive system of longitudinal dunes in the Rub'Al-Khali Desert (the Empty Quarter, Arabian Peninsula), which is the largest sand desert on Earth. Scenes are approximately 90 km (**b**) and 13 km (**c**) wide (at ~*18°33′N* and *48°24′E*). (**d**) Longitudinal dunes trade wind belt of Western Algeria. Image is 40 km wide and at ~*24°21′N* and *4°15′W*. (**e**) Linear dunes in northern Yemen. The image shows a section of 75 km at *16°38′N* and *45°35′E* (Image credit: ©Google earth 2012)

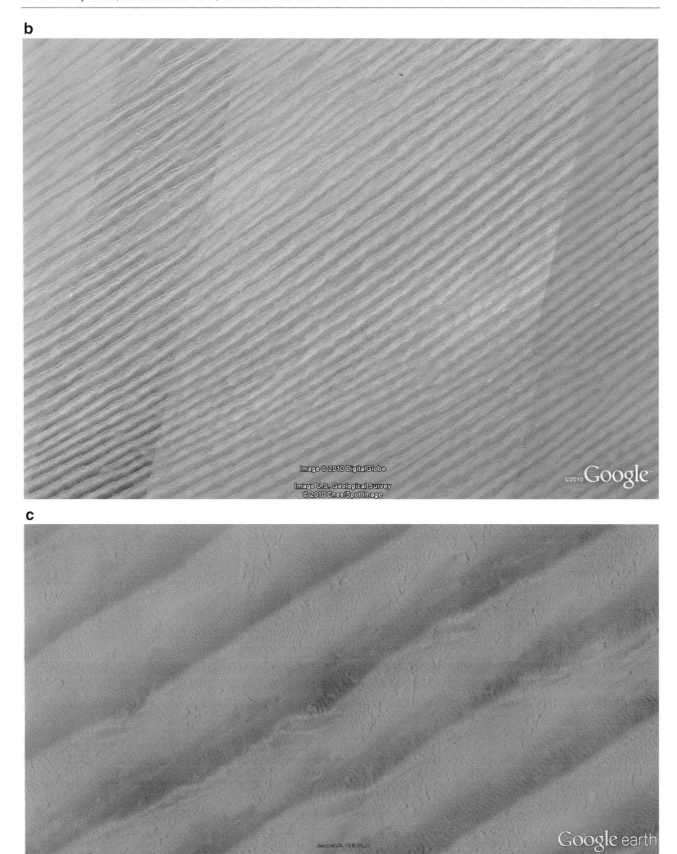

Fig. 11.15 (continued)

d

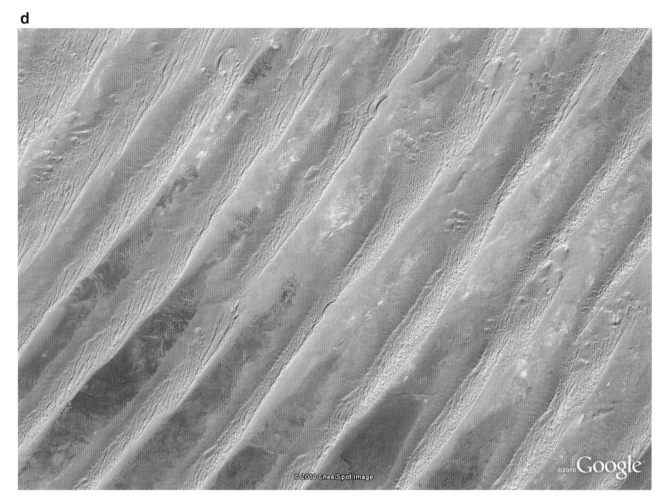

Fig. 11.15 (continued)

e

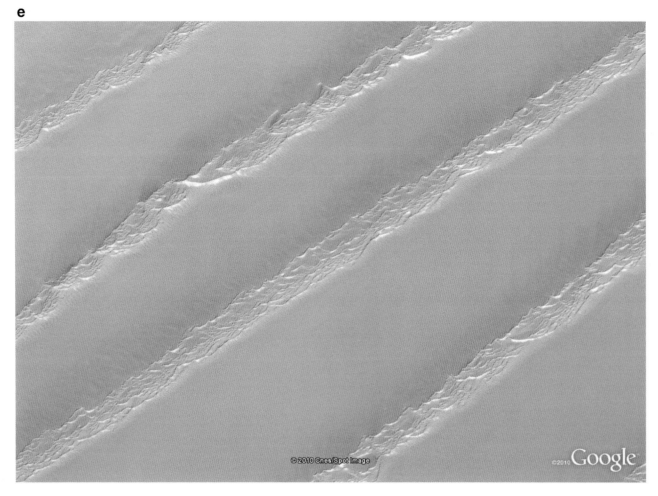

Fig. 11.15 (continued)

Fig. 11.16 (a) A wide area with longitudinal (linear) dunes in the Northern Territory, Australia, at *24°16′S* and *136°15′E*. The section shown is 20 km wide. The *red color* is caused by iron oxides. The dunes, are covered and stabilized by sparse vegetation. (b) A field of narrow linear dunes in the Namib Desert (southwest Africa) at around *23°54′S* and *17°48′E*, on a rocky plateau with salt pans at 1,200 m above sea level. The scene is 34 km wide. Limited sand availability may be responsible for the narrow form of the dunes. (c) Longitudinal or linear dunes in central Australia. Width of image is 42 km wide, at *26°03′S* and *139°47′E* (Image credit: ©Google earth 2012)

c

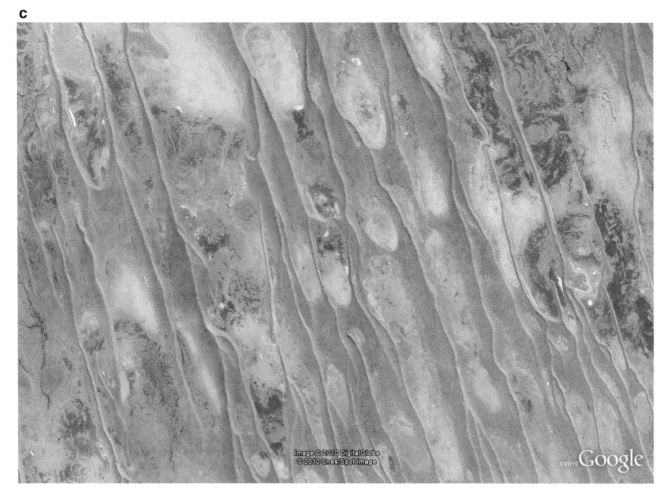

Fig. 11.16 (continued)

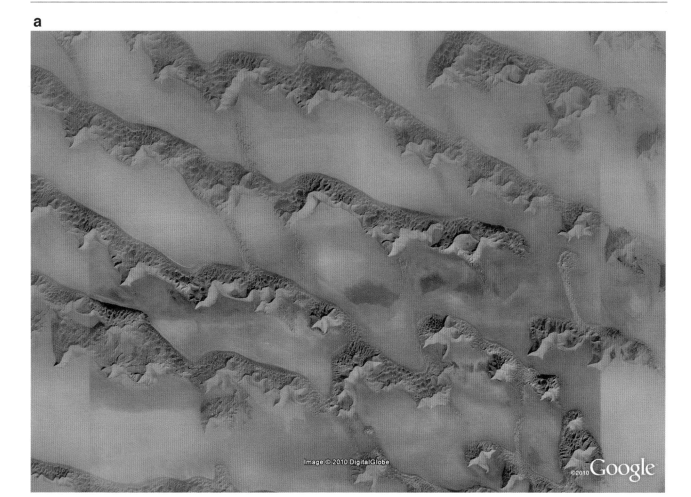

Fig. 11.17 (a) In the easternmost part of Saudi Arabia at around *21°57′N* and *54°09′E*, barchanoid dunes form a network of mega-ripples over more than 30,000 km^2. This image shows a section 70 km wide. Northerly winds have transported the sand in the direction of the Empty Quarter in the Arabian Peninsula. The mega-ripple dunes are more than 100 m high and are decorated by smaller dune forms or even star dunes. (b) Flat-topped mega-ripple dunes in central Mongolia, at around *37°46′N* and *81°45′E* (Image credit: ©Google earth 2012)

b

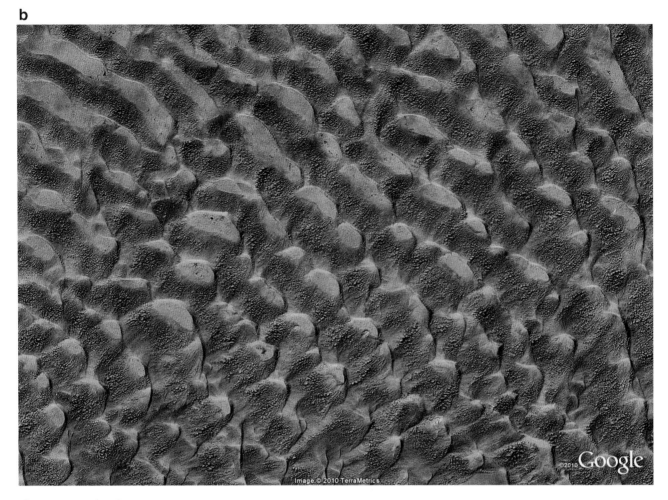

Fig. 11.17 (continued)

Fig. 11.18 Photo of a large sand mountain (star dune) in southern Morocco, at the margin of the large western Erg of the Sahara Desert (Image credit: D. Kelletat)

11 Forms by Wind (or: Aeolian Processes): Deflation and Dunes

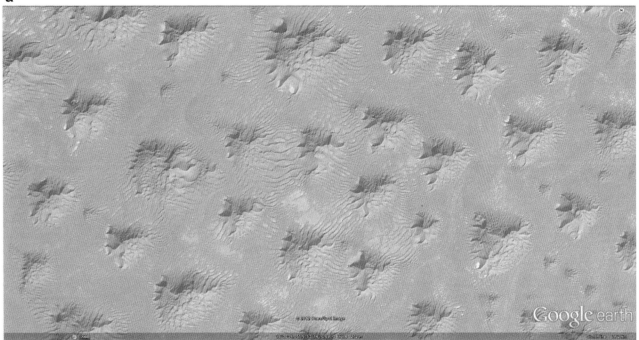

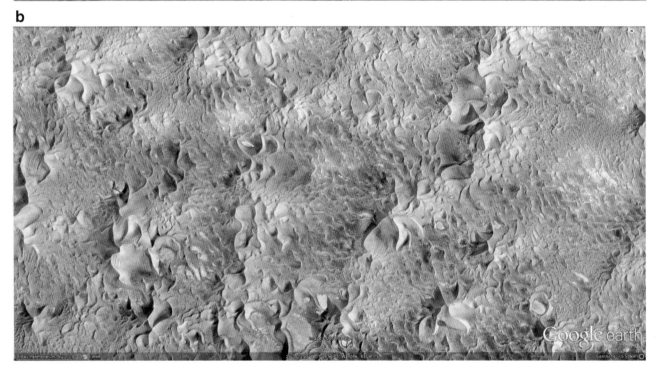

Fig. 11.19 (**a**) In large sandy deserts as in the central Sahara Desert or the southern section of the Arabian Peninsula (western Oman; about *18°31′N* and *53°14′E*), star-like dunes form isolated hills and are the dominant pattern over wide areas. The scene is 7 km wide and the dunes are 50 to >70 m high. (**b**, **c**) In western Algeria at about *29°34′N* and *1°32′W*, sand mountains with star dune character rise from a wide rocky wadi floor. The first high dune crests are nearly 100 m higher than the wadi floor. The scene in (**b**) is about 20 km wide and (**c**) is 4 km wide. The typical star-shaped crests are clearly visible. Closed depressions occur at many places within the dune fields. (**d**, **e**) Sossusvlei region in Namibia, southwest Africa (*24°42′S* and *15°29′E*), where some of the highest dunes on Earth exist. Single dunes may reach more than 300 m in height. The section in (**d**) is more than 40 km wide. The central part of the scene is a wide sand and salt pan; flash floods during torrential rains enter the area from the east. The 5.5 km wide scene in (**e**) shows the curving crests of huge dunes, with star-features and closed depressions in between the highest sand hills (Image credit: ©Google earth 2012)

c

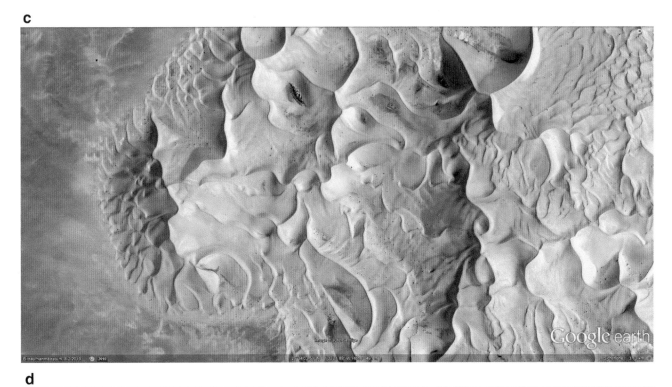

d

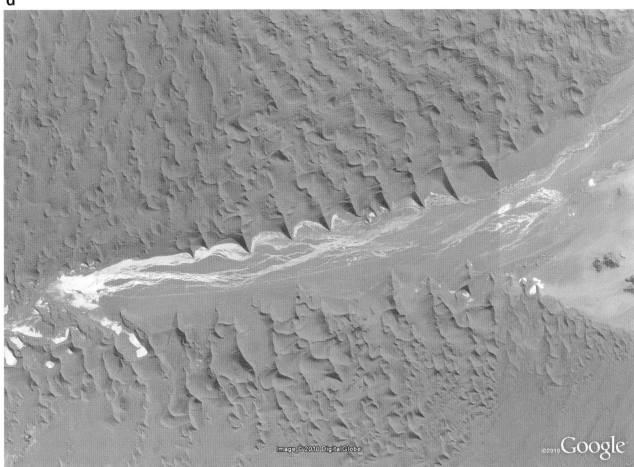

Fig. 11.19 (continued)

e

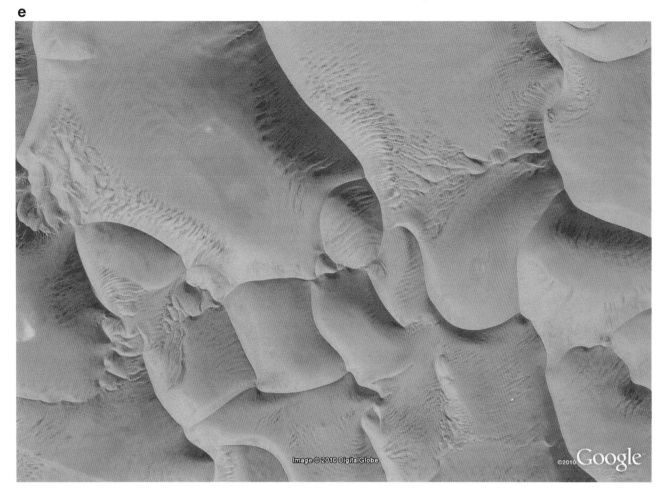

Fig. 11.19 (continued)

needed (e.g., by winds crossing extensive beaches, dry wadi floors and alluvial plains at the foot of mountain ranges, or by eroding geological formations such as sandstones). At places where sand loss exceeds sand input, dunes may abruptly disappear as in the downwind parts of the northeast section of the Namib Desert in southwest Africa (Namibia) where the sand sea immediately stops at the Kuiseb River (Fig. 11.20).

However, where boundary conditions such as regional climate or sand availability significantly changes throughout time, wind-driven sand transport and sand accumulation may be considerably reduced or even stop, and dunes may become stable and vegetated. Fixed dunes (or stabilized dunes) may still characterize the landscape and its geomorphology even when aeolian dynamics receded. Even after several 100,000 years, palaeodunes may still show their original morphological pattern, such as the soil-covered and partly vegetated dunes in Australia's outback and deserts. Investigations on palaeodunes in regards to their stratigraphy and their morphological patterns thus allow us to reconstruct, for instance, past wind systems and sand availability conditions, and ultimately, may give evidence of past climate changes.

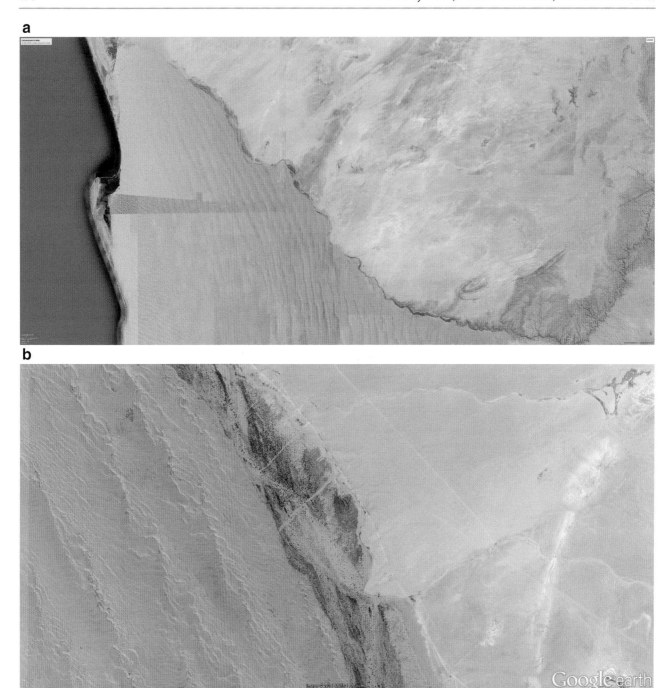

Fig. 11.20 (**a**) The Kuiseb River in southwest Africa forms the abrupt northeast boundary of the large sand sea in the Namib Desert. Scene is about 130 km wide (Image credit: ©Google earth 2012). (**b**) Detail of (**a**) at *23°15′S* and *14°55′E*

Further Readings

Al-Dousari AM, Al-Elaj M, Al-Enezi E, Al-Shareeda A (2009) Origin and characteristics of yardangs in the Um Al-Rimam depressions (N Kuwait). Geomorphology 104(3–4):93–104

Amato J (2000) Dust: a history of the small and the invisible. University of California Press, Berkeley

Andreotti B, Claudin P, Douady S (2002) Selection of dune shapes and velocities. Part 1: dynamics of sand, wind and barchans. Eur Phys J B 28(3):321–339

Arens SM, Jungerius PD, van der Meulen F (2001) Coastal Dunes. In: Warren A, French JR (eds) Habitat conservation: managing the physical environment. Wiley, Chichester, pp 229–272

Arnalds O, Gísladóttir FO, Sigurjonsson H (2001) Sandy deserts of Iceland: an overview. J Arid Environ 47(3):359–371

Arthur SJ (2005) Ventifacts from Alaska to Antarctica, long-term faceting forms these natural wonders. Rock Gem 35(12):80–83

Baas ACW (2008) Challenges in aeolian geomorphology: investigating aeolian streamers. Geomorphology 93(1–2):3–16

Bauer BO (2009) Contemporary research in aeolian geomorphology. Geomorphology 105(1–2):1–5

Besler H (2002) The Great Sand Sea (Egypt) during the late Pleistocene and the Holocene. In: Schmidt KH, Vetter T (eds) Late quaternary geomorphodynamics, Zeitschrift für Geomorphologie Supplementband 127. Borntraeger, Berlin, pp 1–19

Besler H (2008) The Great Sand Sea in Egypt: formation, dynamics and environmental change – a sediment-analytical approach, with contributions from A. Bolten and others. Developments in sedimentology 59. Elsevier, Amsterdam

Besler H, Boedeker O, Bubenzer O (2007) Dunes and megadunes in the southern Namib Erg and the southern Great Sand Sea; a transcontinental comparison. Africa Praehist 21:60–63

Bird ECF (1990) Classification of European dune coasts. In: Bakker TWM, Jungerius PD, Klijn JA (eds) Dunes of the European coasts, Catena Supplement 18. Catena, Cremlingen-Destedt, pp 15–23

Breton C, Lancaster N, Nickling WG (2008) Magnitude and frequency of grain flows on a desert sand dune. Geomorphology 95(3–4):518–523

Bristow CS, Duller GAT, Lancaster N (2007) Age and dynamics of linear dunes in the Namib Desert. Geology 35(6):555–558

Brookes LA (2001) Aeolian erosional lineations in the Libyan Desert, Dakhla Region, Egypt. Geomorphology 39(3–4):189–209

Clemmensen LB, Murray A, Heinemeier J, deJong R (2009) The evolution of Holocene coastal dunefields, Jutland, Denmark: a record of climate change over the past 5000 years. Geomorphology 105(3–4):303–313

Cooke RU, Warren A, Goudie AS (1992) Desert geomorphology. UCL Press, London

Doody JP (ed) (1991) Sand dune inventory of Europe. European Union for Coastal Conservation (EUCC)/Joint Nature Conservation Committee, Peterborough

Eitel B, Bluemel WD (1997) Pans and dunes in the southwestern Kalahari (Namibia): geomorphology and evidence for Quaternary paleoclimates. Z Geomorphol Suppl 111:73–95

El-Sayed MI (2000) The nature and possible origin of mega-dunes in Liwa, Ar Rub' Al Khali, UAE. Sediment Geol 134(3–4):305–330

Fitzsimmons KE (2007) Morphological variability in the linear dunefields of the Strzelecki and Tirari Deserts, Australia. Geomorphology 91(1–2):146–160

Gillies JA, Nickling WG, Tilson M (2009) Ventifacts and wind-abraded rock features in the Taylor Valley, Antarctica. Geomorphology 107(3–4):149–160

Goudie AS (2007) Mega-yardangs: a global analysis. Geography Compass 1(1):65–81

Goudie AS (2008) The history and nature of wind erosion in deserts. Annu Rev Earth Planet Sci 36:97–119

Goudie AS (2009a) Aeolian processes and landforms. Chapter 14. In: Woodward J (ed) The physical geography of the mediterranean. Oxford University Press, Oxford

Goudie AS (2009b) Dust storms: recent developments. J Environ Manage 90(1):89–94

Herrmann J (2006) Aeolian transport and dune formation. Lecture Notes in Physics Series 705. Springer, Berlin/Heidelberg, pp 363–386

Knight J, Burningham H (2001) Formation of bedrock-cut ventifacts and late Holocene coastal zone evolution, County Donegal, Ireland. J Geol 109(5):647–660

Kocurek G, Ewing RC (2005) Aeolian dune field self-organization – implications for the formation of simple versus complex dune-field patterns. Geomorphology 72:94–105

Lancaster N (2007) Dune fields, low latitudes. In: Elias SA (ed) Encyclopedia of quaternary science, vol 2. Elsevier, Amsterdam, pp 626–642

Nordstrom KF, Psuty NP, Carter RWG (eds) (1990) Coastal dunes: processes and geomorphology. Wiley, New York

Pye K, Tsoar H (2009) Aeolian sand and sand Dunes. Springer, Dordrecht

Reffet E, Courrech du Pont C, Hersen P, Douady S (2010) Formation and stability of transverse and longitudinal sand dunes. Geology 38(6):491–494

Seppälä M (2004) Wind as a geomorphic agent in cold climates. Cambridge University Press, Cambridge

Tack F, Robin P (2002) Dunes. Vilo International, Paris

Twidale CR (2008) The study of desert dunes in Australia. Geol Soc Lond Spec Publ 301:215–239

Wolfe SA (2007) Dune fields, high latitudes. In: Elias SA (ed) Encyclopedia of quaternary science, vol 2. Elsevier, Amsterdam, pp 599–607

Wopfner H, Twidale CR (2001) Australian desert dunes: wind rift or depositional origin? Aust J Earth Sci 48:239–244

Glacier Ice and Its Domain

12

Abstract

Glacier ice forms from accumulated old snow, and the erosive power of glaciers flowing down-valley as well as the related production and deposition of large amounts of rock debris are responsible for the large inventory of glacial landforms. Even though we are living in a warm climatic period at the moment (the Holocene interglacial period), glaciers are still common in the cold climates of the higher latitudes as well as in high mountains all over the Earth. There are different types of glaciers depending on the general ice budget, ice temperatures or the general topographical circumstances. Glacier ice may entirely cover large areas as in the case of inland ice and plateau glaciers, or it can be restricted to the mountain relief such as glaciers in cirques, or valley glaciers. Glaciers that exit mountain belts and advance into the foreland are called piedmont glaciers. Glaciers exert enormous erosive forces due to their weight and their ability to transport and use rock debris of different sizes that are frozen at the glacier's base. Rock debris scratches the hard rock at the valley floor and creates typical striae or striations as well as polishing of rock surfaces where finer particles are incorporated in the basal ice. Probably the most prominent example for the smooth forms of glacial erosion and polishing in hard rock surfaces are "*roches moutonnées*", which are characterised by a whaleback-like shape and an asymmetrical long profile. Other well-known forms are cirques particularly in higher mountain areas and U-shaped cross profiles of valleys which have been eroded by former glaciers. In turn, moraine ridges consisting of glacial debris give depositional evidence of former glaciations and deglaciations. Large amounts of debris material are transported on the surface, within the glacier, at its bottom, or along its lateral boundaries as "*moraines*". The sediments of these moraines are called till. After deglaciation, remnants of the lateral deposits remain in the form of long and sharp-crested ridges (lateral moraines) and large amounts of debris that were pushed in front of the former glacier are left to form a terminal moraine. Other forms are created below the moving ice such as "*drumlins*", or as relics of subglacial meltwater streams below the glacier such as the long and narrow "*eskers*" or "*osers*".

In contrast to other planets in our solar system, the mean temperature (15 °C) and the temperature range on Earth allow water to occur both in solid and liquid form, and the elements of water (oxygen and hydrogen) are available in our atmosphere in large amounts. During the geological history of our planet, ice has repeatedly covered large parts of the Earth's surface. Ice ages occurred at least five times during the geologic past of our planet – the first and oldest in the Proterozoic period, the second (and largest, the so-called "Snowball Earth" with ice assumed to have reached the equator) in late Precambrian times ~850–630 million years ago, the third in late Ordovician and Silurian times (~460–420 million years ago), and the fourth in the Carboniferous and early Permian, ~360–260 million years ago. The fifth and last ice age started ~2.4 million years ago at the beginning of the Quaternary. During the Pleistocene (as part of the Quaternary, 2.4 million to ~12,000 years ago), remarkable climatic fluctuations caused recurring and alternating periods of colder and warmer conditions, with differences of the mean temperature between cold and warm periods up to ±10 °C. In combination with significant changes in precipitation, these climatic fluctuations led to the worldwide recurrent advance and recession (i.e. melting) of glaciers. In particular, higher latitudes were more affected by these fluctuations. In the glacial periods, thick ice sheets formed (e.g., in Scandinavia and North America but also on top of mountain belts), and the inland glaciers of Scandinavia reached the northern parts of central Germany. During interglacial periods, however, glaciers melted and retreated, and warm climatic conditions comparable to those we experience today occurred between the cold periods. Moreover, the formation and disappearance of glaciers had direct effects to the oscillations of the global sea level of more than 100 m due to the large amounts of water trapped in the glaciers during glacial periods. While the last glacial period (called Weichsel or Würm in central Europe, or Wisconsin in North America) still belongs to the Pleistocene, the current warm climatic period started ~12,000 years ago and is still ongoing. Apparently, these fluctuations had a strong influence on all kinds of environmental conditions including the biosphere and, ultimately, influenced the distribution of man. In fact, the fluctuations of glaciers and ice sheets had a remarkable impact on the geomorphology of the affected regions, and wide landscapes were formed by the erosive force of glacier ice and ice-related deposition.

To understand the inventory of glacial geomorphology and deposits allows researchers to infer past climate changes; however, also the ice itself serves as an important archive for the reconstruction of past climates. During the annual deposition of snow, air bubbles and fine particles such as dust, volcanic ash, or pollen as well as chemical elements are trapped in the ice (Figs. 12.1a, b and 12.2), and their analyses give insights into palaeoclimatic and palaeoenvironmental conditions. In Antarctica, an ice core of more than 3,000 m depth reaches back in time to more than 700,000 years ago. The history of ice deposition may particularly be seen in areas where glaciers and active volcanoes exist in close vicinity,

Fig. 12.1 (a) A series of different snowfalls from one winter is preserved on an alpine pass. Each darker stratum represents a phase of melting, where snow disappeared but other particles have been enriched within the remnants of the old snow. Similar records are developed in a same manner in glacial ice over much longer time periods (Image credit: D. Kelletat). (b) This 450 m long glacier at Gerenpass in the Swiss Alps (*46°30'05.46"N, 8°26'51.84"E*) exhibits many dirt layers in its body of ice, which represent the many former summer periods as dust and dirt within the snow are compacted into glacial ice (Image credit: ©Google earth 2012)

b

Fig. 12.1 (continued)

Fig. 12.2 Ice crystals from an alpine valley glacier, grown together with a diameter of up to more than 10 cm each. White tiny spots in the left upper part of the image are small bubbles of air in the ice (Image credit: D. Kelletat)

such as on Iceland. There, volcanic eruptions may leave dark ash layers on the snow, which is later transformed into glacial ice. The dark ash layers appear in the lower areas of these glaciers with dominant ablation such as the glacier tongue, and beautiful patterns of clean ice and ash-bearing ice can be observed on the glacier surface (Figs. 12.3 and 12.4).

Glaciers are formed and sustained by the accumulation of snow. In general, glaciers may develop in areas with a positive balance of snow accumulation, where snow is left at the end of the summer (i.e., at the end of the ablation period) year by year and accumulates for decades and centuries. The small and fragile snow crystals are transformed into glacial ice with a density of ~0.9 g/cm^3 by the overlying pressure, recurrent melting and freezing, and nearly all of the air has been pressed out (Figs. 12.1 and 12.2). In temperate regions like the European Alps or North American Rocky Mountains, where meltwater percolates through the snow and recrystallizes, it takes about 20 m thickness of snow and two to three decades to transform the snow into large irregular ice crystals (up to more than 10 cm across), grown together with temperatures just below 0 °C. In very cold areas like in Antarctica, however, with the pressure of the thin snow cover from year to year, this process may need about 100 years and nearly 50–100 m of snow to produce ice crystals that are only about 1 cm wide. This ice is cold as the snow from which it has formed, −20 to −30 °C or even colder in the central parts of Antarctica. On the other hand, glaciers loose ice either by melting (ablation), or by calving into lakes or even the oceans, where the ice blocks form icebergs (such as in the Arctic and Antarctic) (Figs. 12.5, 12.6 and 12.7).

We can categorize glaciers based on their relation to the surrounding relief (Figs. 12.8, 12.10 and 12.11). Cirque (or Kar) glaciers are generally restricted to the semi-enclosed cirques, which also represent the places where they started to form. Valley glaciers may fill a valley and are elongated like a river of ice (Figs. 12.11 and 12.12). Both types of glaciers are adapted to existing configurations of the relief, though in turn changing the relief as well. Piedmont glaciers are glaciers which exit mountain belts via valleys and spread in front of the mountain belts forming broad lobes in many

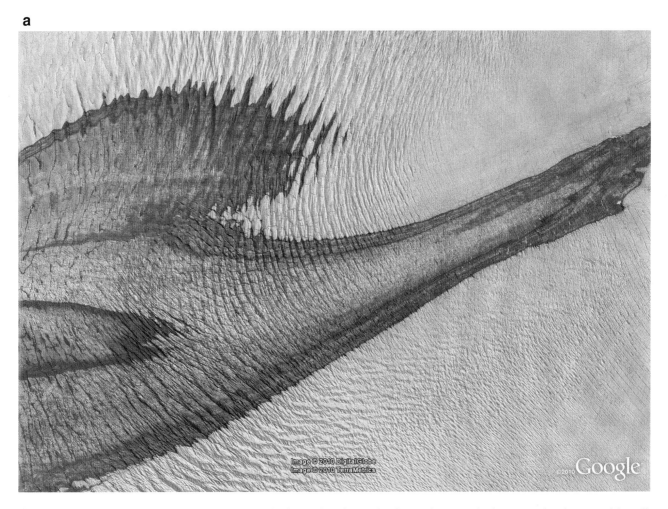

Fig. 12.3 (a) Dirt layers on this Chilean glacier are particles from volcanic eruptions, like tuff, ash and cinder. These may derive from a single eruption with different rhythms and therefore show material of different grain size. (b) In this Chilean glacier tongue the dirt layers from volcanic eruptions are clearly separated and preserved in well-defined layers even though they have been transported from the highest parts of the glacier (fallen on snow) for at least 40 km (Image credit: ©Google earth 2012)

b

Fig. 12.3 (continued)

a

Fig. 12.4 (**a**, **b**) In Iceland, where highly active volcanoes are covered by glaciers, the mixing of tephra and snow occurs in short time scales, sometimes several times within 1 year. Therefore, we see multiple dark ash layers of different intensity. (**c**) In the northern part of Vatnajökull, the largest Icelandic glacier, this tongue exhibits a very high number of ash layers, giving a good record of volcanic events for most probably several hundred years of time (Image credit: ©Google earth 2012)

b

Fig. 12.4 (continued)

c

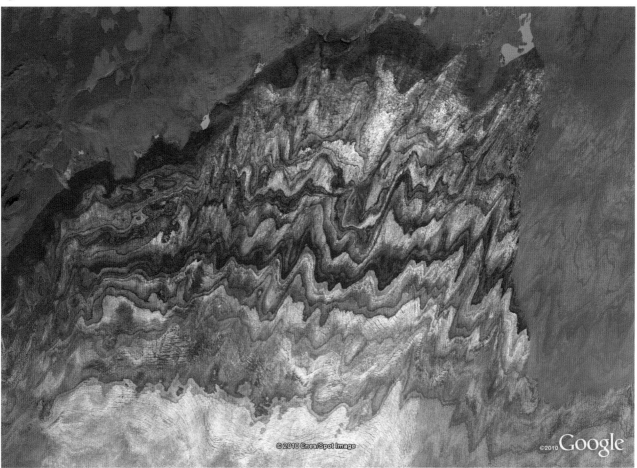

Fig. 12.4 (continued)

12 Glacier Ice and Its Domain

a

Fig. 12.5 (a) The famous Perito Moreno Glacier in southern Patagonia (Argentina) at *50°28'S* and *73°03'W* with a tongue nearly 3 km wide. The glacier oscillates within several years and closes the eastern section of Lago Argentino. As a consequence, lake water is stored up to about 20 m higher than in the northern part, and the glacial tongue may float and break. From time to time, the southern lake bursts out. The satellite image shows a small barren bank along the lake on the right part of the image, which marks the maximum high water line without vegetation (Image credit: ©Google earth 2012). (b) Northern part of the tongue of the Perito Moreno Glacier calving into Lago Argentino (Image credit: D. Kelletat)

Fig. 12.5 (continued)

Fig. 12.6 (a) San Quintin Glacier (southern Chile) calving into its terminal lake. (b, c) Glaciers from the Southern Patagonian Ice Field in Chile may reach the ocean in fjords, with steep fronts and sharp crevasses and calving icebergs. Scenes are several 100 m wide (Image credit: ©Google earth 2012)

Fig. 12.6 (continued)

c

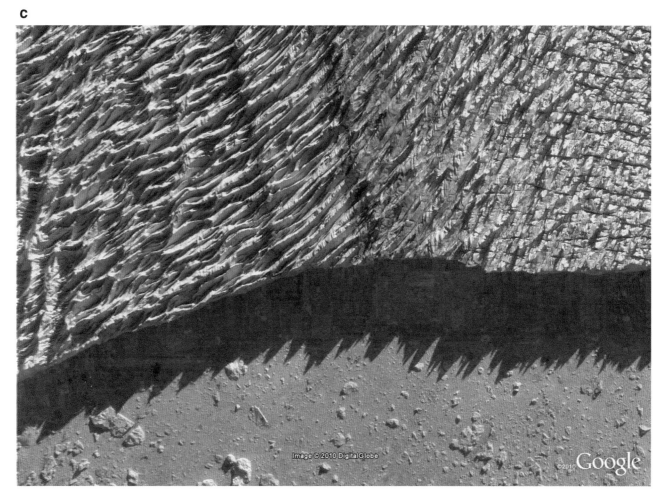

Fig. 12.6 (continued)

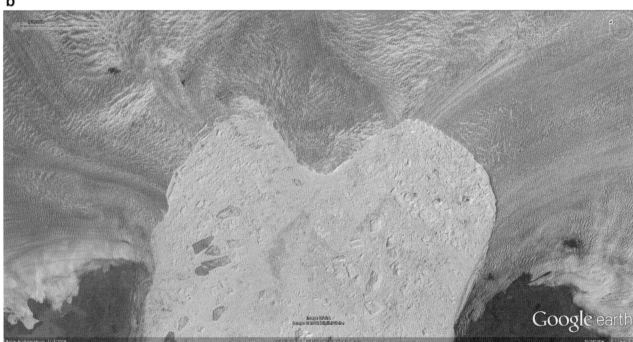

Fig. 12.7 (a) The grey ice in the background is the Jacobshavn Glacier, the most productive outlet glacier in the northern hemisphere (western Greenland, near Disko Bay/Ilulissat). Icebergs (the lighter colors) are calving constantly from its front, filling Jacobshavn Isfjord with thousands of icebergs in a length of about 60 km and a width of about 6 km (Image credit: D. Kelletat). (b) Same area as seen from Google Earth (Image credit: ©Google earth 2012)

Fig. 12.8 (**a**) A plateau glacier entirely covering the underlying relief. This glacier is part of the Columbia Icefield in the Canadian Rocky Mountains (Image credit: D. Kelletat). (**b**) The largest European glacier, the Vatnajökull in Iceland, covers several active volcanoes and represents a typical plateau glacier (Image credit: ©Google earth 2012). (**c**) Glaciers may also develop at very steep slopes frozen to the rock (Mt. Cook on New Zealand's South Island) (Image credit: D. Kelletat)

Fig. 12.8 (continued)

cases (Fig. 12.10a–d). In contrast, plateau glaciers may completely cover entire mountain belts, and the underlying rocky relief may be entirely invisible. Combined valley glaciers form a network of ice streams. Finally, inland ice is the term for the largest continuous ice masses such as those covering Greenland or Antarctica. During the glacial periods of the Pleistocene, inland ice masses covered large parts of northern Europe, northern Asia and northern America, and the total ice mass on our globe was about three times higher compared to the present situation. Glaciers in high latitudes may have reached the ocean and formed floating shelf ice as presently in Antarctica.

Glacier ice behaves in a viscose or plastic manner when moving or "creeping" downslope. However, the temperature of a glacier also influences the ice movement. Temperate glaciers react more plastically and show typical forms of flowing, as shown in many of the following figures (Figs. 12.9, 12.10, 12.11 and 12.12), whereas ice of cold glaciers often breaks into separate parts and then moves as a stream of compact blocks down-valley. The surface of these cold glaciers is rough and shows a lot of deep fractures and ice towers (*"séracs"*), especially when crossing topographical steps. The surface of a temperate glacier is usually smoother, and fractures usually close again immediately below topographical steps in the valley (Fig. 12.9a, b). The difference between cold and temperate glaciers may also be expressed in the behavior of medial moraines, or *"ogives"* (infills of light fresh snow or darker debris in crevasses formed over rocky steps) and their deformation by the glacier's movement (Fig. 12.12c–e). Here, fissures form over steps in a valley or where friction along the valley sides breaks the ice.

Generally, temperate glaciers contain large amounts of meltwater, produced by insolation at the surface as well as by pressure and slightly elevated temperatures at the glacier's base, or within caverns and fissures that are mostly at the interface of the glacier and the rock. The meltwater flows under the ice towards the end of the glacier; long tunnels in the ice are formed, and the subglacial water appears as a meltwater river at the glacier mouth (Fig. 12.13). However, meltwater creeks or rivers may be observed even on the surface of the Greenland ice dome. Where disappearing into fractures of the glacier, deep vertical shafts called "glacier mills" are formed.

The Antarctic ice sheet can be up to 4,000 m thick in some places. As the ice pressure leads to ice melting at its base large meltwater lakes can form. It is easy to understand

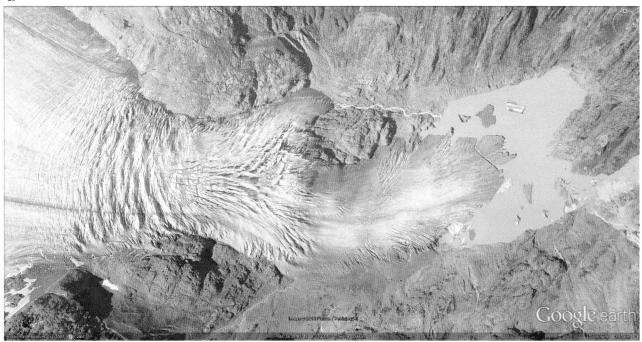

Fig. 12.9 (a) The Gauli Glacier in Switzerland. Its fine pattern of crevasses reflects the topography over which it flows. Over steeper valley areas, the fractures are larger. A second pattern of fractures is caused by the friction of the moving ice along the valley slopes (Image credit: ©Google earth 2012). (b) A glacier cascading down in Alaska's Glacier Bay National Park, north of Juneau (Image credit: D. Kelletat). (c) Cirques with remnant glaciers, separated by sills from the lower valleys in the Swiss Alps (Image credit: ©Google earth 2012)

Fig. 12.9 (continued)

Fig. 12.10 (a) A piedmont glacier in the "Dry Valleys" of Antarctica (Image credit: J.-F. Venzke). (b) Axel Heiberg piedmont glaciers (Canada) at about *79°42′N* and *74°51′W*. Diameter of the glacier lobe is 1 km and more (Image credit: ©Google earth 2012). (c) Aspect of the Malaspina Glacier from an airplane at about 5,000 m above sea level (Image credit: D. Kelletat). The Malaspina Glacier forms a 5,000 km² large piedmont tongue in the Alaskan Panhandle. The concentric pattern is most likely due to pressure phenomena at the outer ice front pushing towards the terminal moraine. Similar patterns of moraine deformation on and in glacier ice may occur during glacier surges or where obstacles force the glacier to unusual pathways (d) The Malaspina Glacier from space (Image credit: ©Google earth 2012)

Fig. 12.10 (continued)

d

Fig. 12.10 (continued)

Fig. 12.11 A network of ice streams (valley glaciers) in eastern Greenland in a 77 km wide section. Center of the image is at about *66°32′N* and *36°24′W*. At the confluence of two valley glaciers, their lateral moraines are combined to form a medial moraine, which follows the glaciers as a significant debris line in the center of the glacier downvalley. This is a type of laminar movement without significant turbulence in the glacier. The number of these surface moraine stripes at any cross section of a large glacier system points to the number of glaciers that have fed into the main glacier. The main glacier on this image is 3 km wide (Image credit: ©Google earth 2012)

12 Glacier Ice and Its Domain

Fig. 12.12 (**a**) A Greenland glacier blocks a crossing valley and flows for a short distance into two directions. (**b**) Himalayan glacier overflowing a valley step and therefore breaking across along its entire width. (**c, d**) The ogives show the intensity of deformation on a moving valley glacier (Mer de Glace, French Alps) and at the same time illustrate a special flow character: evidently slower movement along the banks because of friction, and fastest in the center, which is furthest away from lateral friction and has the maximum ice depth. (**e**) Ogives on a fast flowing glacier in the Karakorum Mountains (Image credit: ©Google earth 2012)

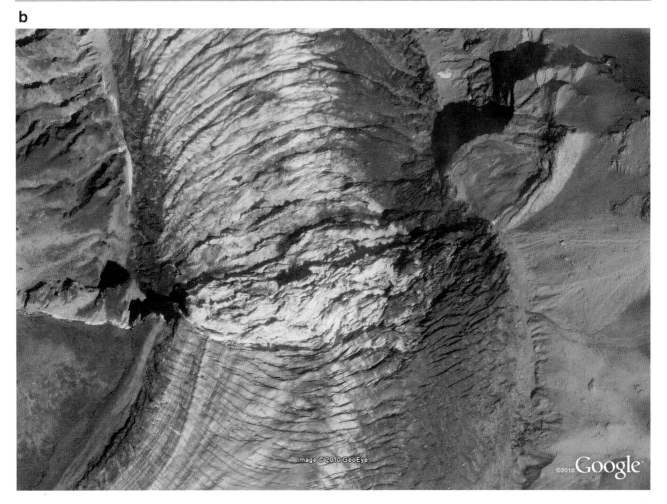

Fig. 12.12 (continued)

c

Fig. 12.12 (continued)

d

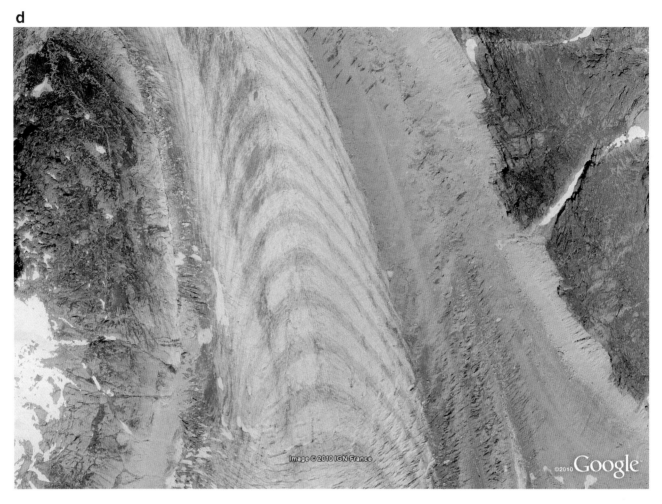

Fig. 12.12 (continued)

e

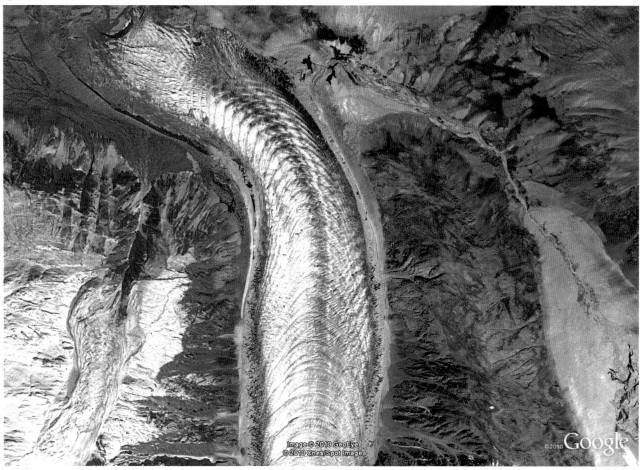

Fig. 12.12 (continued)

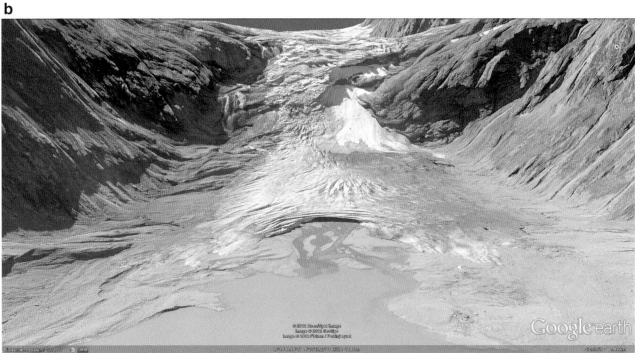

Fig. 12.13 (a) At the mouth of a glacier, meltwater may appear in a wide tunnel, formed by the running meltwater. Snout of Nigardsbreen in Norway (Image credit: D. Kelletat). (b) Glacier snout of the Trift Glacier (Swiss Alps) with an adjacent meltwater lake (*46°40′N, 8°21′E*) (Image credit ©Google earth 2012)

that a meltwater film at the base of a glacier will enhance its ability to move and flow over the rocky valley floor. In general, the velocity of glacial movement is dependent on the temperature, the mass pushing the ice forward, ground friction, the slope angle, and of course the ice budget. If the ice budget is positive (i.e., a lot of snow accumulates and is transformed into ice), the nourishment is positive and there is a considerable glacial movement. If nourishment is low (i.e., a small or no surplus of snow at the end of the summer), the movement will be minimal. In the European Alps temperate glaciers of some kilometers in length will move about 30–60 m/year. In the steep New Zealand Alps, longer glaciers move more than 100 m/year. And in Greenland, the outlet glaciers may reach velocities of up to 30 m/day which is about 10 km/year in valleys fed from the large inland ice ("outlet glaciers"). In cases of extreme nourishment by snow or ice avalanches (often triggered by earthquakes as in Alaska, USA), a glacier may rush forward in a surge of several kilometers in only weeks or months. However, the processes determining the recession or advance of glaciers are rather complex, and changes to a number of both local and global conditions may be of more importance for glaciation and deglaciation (Fig. 12.14) than mean temperature or precipitation.

Rock fragments and all kinds of rock debris may accumulate on the surface of cirque glaciers, valley glaciers and all other types which are embedded to a relief. Rock fragments are also broken down by the glacial ice from the valley sides and the valley floor. While fine particles and small stones on a glacier will sink into the ice because they are heated up by insolation, large boulders may protect the glacier surface from insolation: the ice melts away next to the boulders, but is preserved below the boulder forming a "glacier table" (Fig. 12.15). Debris may become incorporated into the glacier's interior or it can be washed into ice fractures and fissures by meltwater. All the debris masses accumulating on top of, within, below and along the ice are subsumed under the term "moraine" or with the geologic term "till" (also "drift" or "glacial drift"). At the confluence of two glaciers, the moraines at the side of the glaciers, lateral or side moraines, are combined to form medial moraines, which are visible mostly as parallel stripes of all kinds of rock debris (Fig. 12.16). This illustrates the quasi-laminar flow behavior of ice without turbulent mixing like flowing water in rivers. However, the deformation also depends on the topographic conditions and on lateral friction. The rock debris transported by the glacier may finally be pushed and pressed together by the ice front into a terminal moraine – these terminal moraines may give morphological evidence of former glacier positions, and series of moraines are good indicators of glacier oscillations.

In very high and steep mountainous regions like the Himalayas, valley glaciers develop under special conditions. Here, the area of nourishment is very steep in most cases and snow accumulation on top of the glacial ice is mainly due to snow and ice avalanches. These processes transport a lot of rock debris on top of and into the glacial ice, and more debris is added along the often very long glaciers (>40–70 km). However, their length may also result from the fact that the surrounding high mountains receive a lot of snow from monsoonal precipitation. The steep and deeply incised valleys protect the ice from insolation due to shadowing, and the thick layers of moraine material in the lower parts of the glaciers protect the ice sheets as well (Fig. 12.17a, b). In such cases, glaciers may reach into rather warm altitudes of lower elevation. Sometimes the glacier tongues just consist of "dead" or stagnant ice, which has no connection with the actively moving glaciers. Numerous ponds of meltwater and collapsed stagnant ice or kettle holes give evidence for the presence and slow melting of ice under the moraine layers (Fig. 12.18a, b).

The formation of a glacier may start in small depressions (a hollow or trough) in the highest parts of a mountain belt. When climate changes to cooler conditions with a lot of precipitation, snow successively accumulates in these depressions, and it will be present over a longer period during the year. At the boundary between the white snow or ice and the darker underlying rock, frost shattering is very active and delivers a lot of debris – the area of the initial depression increases and steepens backwards. Ongoing snow accumulation and ice formation may increase the thickness of the ice body, and at some point in time it will slowly start to move down-valley and carve into the rock, transporting debris downwards as it moves. Meltwater will additionally transport the fragments of frost weathering and the initial depression is transferred into a cirque or semi-cirque. Cirques may have a flat bottom, but in many cases a sill separates the cirque and the slopes below the cirque (Figs. 12.9c and 12.19). In the European Alps, these cirques are called *"kars"* where the former glaciers have disappeared during the warm period of the Holocene; however, numerous cirque glaciers are still present, and in many cases they are source areas of valley glaciers.

Moving glaciers are very powerful agents for the formation of landscapes. Where rock debris at the base of moving glaciers scratch rock surfaces, so-called *"striae"* or *"striations"* (i.e., stripes, see Fig. 12.20) are carved into the rock. This evidence gives information about the presence of former glaciers, and the orientation of striations allows for inferring directions of ice movement (detersion). Hard rock may be polished where fine particles such as silt is present between the moving ice and the basal rock (Fig. 12.20). Meltwater flowing under high hydrostatic pressure is also capable of polishing rock surfaces with its suspension load. Carved signatures of this flowing process such as whirlpools, potholes or depressions of different forms are common as well (Fig. 12.20a, b). Moreover, at the contact of glacial ice

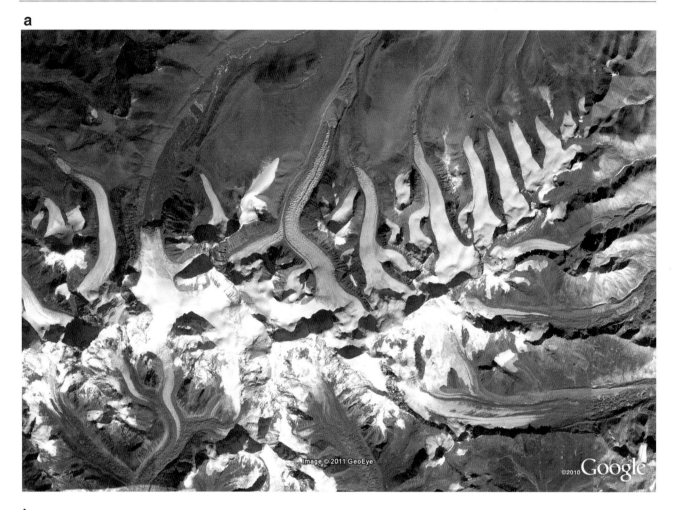

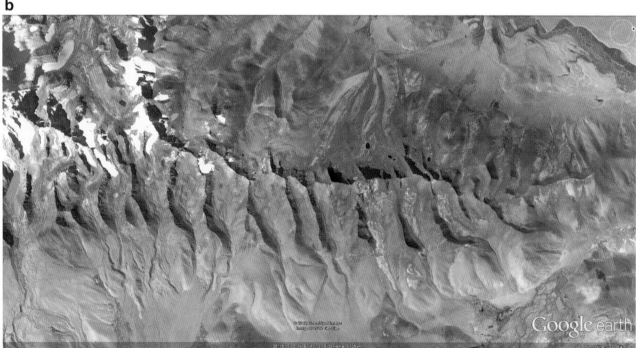

Fig. 12.14 (a) Glaciers in southern China with only minor signs of a younger glacier recession (i.e., glacier melting phase). Scene is at *29°1′N* and *83°3′E* with a width of 35 km. (b) Former glaciated mountain chain, now free of ice, but with typical U-shaped valleys and moraines reaching into the basin south of the mountain range. *Left*: remains of glaciation are visible (Image credit: ©Google earth 2012)

Fig. 12.15 A large boulder has protected the glacial ice beneath from radiation and melting and formed a "glacier table" in the European Alps (Image credit: D. Kelletat)

a

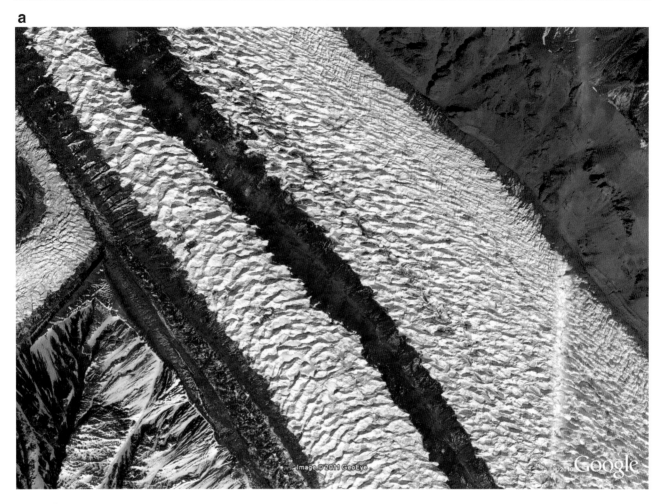

Fig. 12.16 Glaciers with medial moraine bands, documenting the confluence of a few or even many separated ice streams. (**a**) Example from the Himalayas, and (**b**) from eastern Greenland (Image credit: ©Google earth 2012)

b

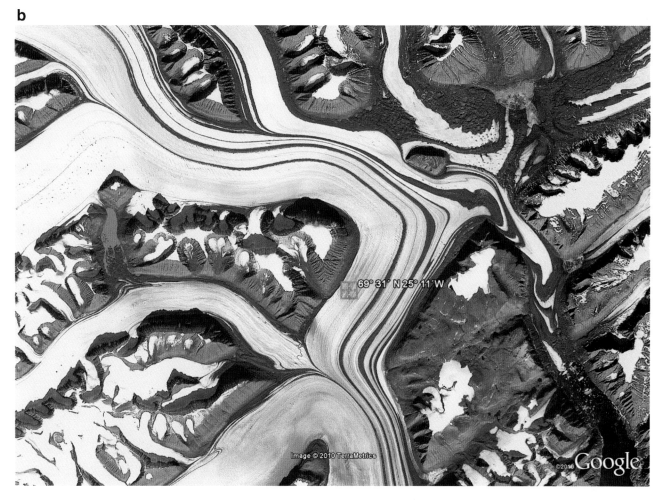

Fig. 12.16 (continued)

a

Fig. 12.17 (**a**, **b**) Two examples of Himalayan glaciers with a thick cover of surface moraines protecting the glacier ice. As a result, the glaciers end with thick tongues, which may also point to a period of glacier advance (Image credit: ©Google earth 2012)

b

Fig. 12.17 (continued)

a

Fig. 12.18 (**a, b**) Southern Asian glaciers in the Himalayas, where the ice is almost entirely hidden under thick moraine material, but sinkhole-like depressions show that it still exists below the till (Image credit: ©Google earth 2012)

12 Glacier Ice and Its Domain

b

Fig. 12.18 (continued)

Fig. 12.19 (**a**) Large cirques in the southern Canadian Rocky Mountains (Image credit: D. Kelletat). (**b**) A group of cirques in northern Iceland. The dark patches are lakes and document that the cirques were deepened by glacial erosion, now separated from the slopes of the valley by sills (Image credit: ©Google earth 2012)

Fig. 12.19 (continued)

Fig. 12.20 (a) Polishing on a freshly exposed rock in the Swiss Alps. Glacier movement was from the left, where the rock rises smoothly. The steep leeward sides (*right*) are due to glacial quarrying. (b) This perfect polishing on gneiss rocks in western Greenland is supported by fine silt and clay particles from meltwaters (Image credit: A. Scheffers). (c) A roche moutonnée on amphibolite near Kangerlussuaq in western Greenland showing fine striations and polishing (Image credit: D. Kelletat). (d) Roches moutonnées from the Swiss Alps (Image credit: A. Scheffers). (e, f) 50 km wide sections of roches moutonnées in northern Canada at *68°* to *69°N*. Ice flow was from northwest in (e), and from the north in (f) (Image credit: ©Google earth 2012)

Fig. 12.20 (continued)

d

e

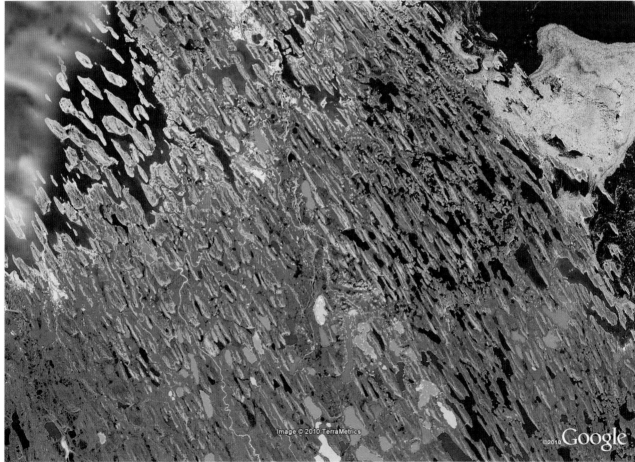

Fig. 12.20 (continued)

f

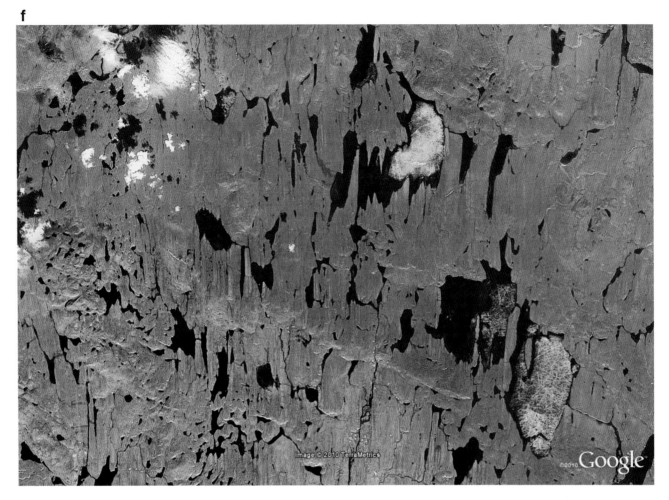

Fig. 12.20 (continued)

and the hard rock, rock fragments are broken up or plucked out of the surface (glacial quarrying). Large or very resistant rock fragments trapped in the moving ice may be blocked by obstacles like hard, resistant minerals or a fracture in the rock basement. The pressure at this point slowly builds up until the rock fails and a shell-shaped fragment is broken out. This leaves a sickle-like fracture in the rock, called "*muschelbruch*" in German (see Fig. 12.21).

Where the rocky basement is riddled with fractures or joints, a swarm of whaleback-shaped rocky hills may form by glacial quarrying, with the orientation of their elongation showing the flow direction of the glacier. These typical rocks are characterized by a lower inclination to the exposed side and a steeper slope with glacial quarrying at the leeward side. Here, pressure and basal meltwater is reduced, and the basal glacier ice rips out rock fragments while the air between ice and rock promote further frost shattering. These whaleback-shaped rocks are called "*roches moutonnées*" (French, meaning "backs of sheep"; Fig. 12.20). They form at places with significant ice pressure (mostly at the bottom of a valley), but also up to the lateral ice boundaries. Striations, polished surfaces and roches moutonnées as documents of glacial erosion make it quite easy to reconstruct the thickness of past ice flow in mountain valleys, even if no glaciers are present today. Multiple glaciations may leave landscapes characterized by overall smooth forms (Figs. 12.22 and 12.23); in contrast to the so-called "*nunatakkers*" (derived from the Inuit language of Greenland, see Fig. 12.23), which are mountains that did not experience glacial erosion. These parts of the mountain belt have been free of ice for a long time and they show the typical sharp forms of areas with dominant frost shattering.

Where glaciers flow down a valley, the form of the valley (i.e. its former pre-glacier cross section of fluvial origin, such as a shallow trough, a V-shape valley, or even a gorge) will be transformed into a wider U-shaped cross profile (glacier carved valley, glacial trough, U-shaped valley). These profiles are typical for nearly all mountain areas which have been covered by glaciers during the glacial periods in the

Fig. 12.21 "Muschelbrüche" on roches moutonnées are developed where shell-shaped flakes of schists have been broken out in the Austrian Alps. Movement was from right to left (Image credit: D. Kelletat)

Fig. 12.22 The landscape in upper Yosemite National Park of California, USA, shows huge roches moutonnées-like mountains and an overall smooth and round appearance of the rock surface, caused by a thick ice cover during several glacial periods of the Pleistocene (Image credit: D. Kelletat)

Fig. 12.23 In central Switzerland (Aare district), we can observe the uppermost limit of ice polishing during the Pleistocene glacial periods, and the line of nunatakker marking areas that have not been covered and rounded by glacier ice (Image credit: D. Kelletat)

Pleistocene (Fig. 12.24). The preservation of this typical U-shaped profile after a glacier receded, however, depends on the resistance of the rocks, mass movements at the steep slopes, and the amount of valley infill with fluvial deposits. Glaciers may also carve out loose material to create elongated finger lakes or glacier lobe lakes (Fig. 12.25).

As glaciers are very active eroding agents, they produce a lot of rock debris of all grain sizes. We have learned that either the glaciers themselves transport the debris (i.e., in the form of moraines) or their meltwater does. However, at places where the motion of the glacier and thus the transport of the rock debris ceases (e.g. when a glacier melts down after having advanced in the past), the loose material is deposited and landforms of different sizes and shapes are created. The particular geomorphology of these depositional landforms gives evidence of the deposits' origin and may also store information about the timing of the last transport of the material. A typical ground or basal moraine forms at the base of a former glacier from the debris left during its melting/recession. It consists of a mixture of fine and coarse particles usually in all shapes and sizes (well-rounded, partly rounded, angular, silt, sand, pebbles, and large boulders) (Fig. 12.26). Ground moraine landscapes are ones with a rather hilly topography and an irregular pattern of both depositional and erosional features. A terminal moraine develops where debris material at the glacier's front is pushed down-valley, creating a wall of debris material in front of the glacier (Fig. 12.27). It may be composed of the same material as the ground moraines, but it may also contain larger amounts of angular fragments from the surface or from inside the glacier, where no erosion or rounding of rock material occurred. Terminal moraines may survive long time periods as particular landforms and since they reflect the outer limit of glacier ice during a certain point in time, they can provide a record of the former extents of previous glacier advances. A series of terminal moraines for instance, in northern Germany, recorded the complex history of the recurring advance and melting of the Scandinavian inland ice reaching central Europe. Finally, lateral moraines may be preserved in mountain areas, but their preservation in the landscape generally is less common over longer time periods (Fig. 12.27).

If loose moraine material is again overrun by a glacier, whaleback-like elongated hills of glacial deposits can develop. However, in contrast to the roches moutonnées composed of hard rock, their steeper sides are directed towards the former glacier. These "*drumlins*" (derived from

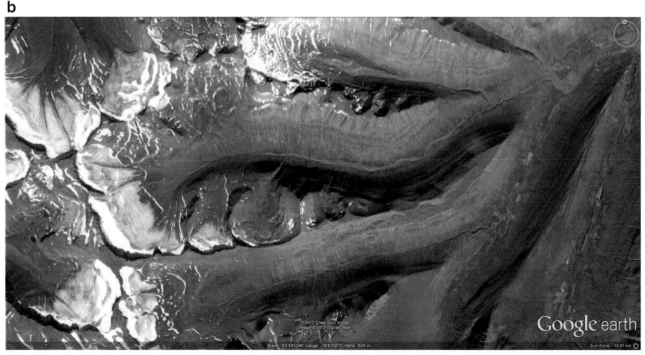

Fig. 12.24 (a) Former glaciated cirque and U-shaped valley dominated landscape in eastern Russia at about *61°37′N, 172°14′E*. (b) U-shaped valleys in Iceland at about *65°38′N, 18°36′W*. (c) This image shows a typical U-valley in the valley of the Krimmler Ache, Austrian Alps at *47°6′N* and *12°11′E* (Image credit: ©Google earth 2012)

Fig. 12.24 (continued)

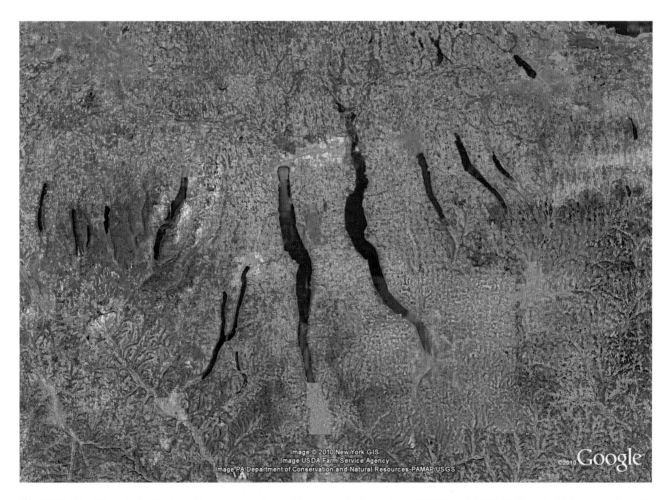

Fig. 12.25 The "Finger Lakes" (up to 70 km long) are places of former glacier tongues during glacial periods that carved out these depressions into a flat landscape east of the Great Lakes of USA and Canada. The site is at about *42°41′N* and *76°48′W*. The glacial ice came from the north (Image credit: ©Google earth 2012)

12 Glacier Ice and Its Domain

Fig. 12.26 (**a**) Chaotic ground moraine particles (till) from a valley bottom in southern Switzerland. Most of the deposit has been washed away, but under large boulders columns or needles of the sediment may be preserved. They are called "earth pyramids" in the European Alps, although they mostly have the form of columns (Image credit: D. Kelletat). (**b**) Deformation of meltwater deposits by an advancing glacier east of Bariloche in western Argentina about 13,000 years ago at the end of the last glaciation. (**c**) Detail of (**b**) showing the small faults in the silt strata documenting that the sediment was frozen during the last movement (Image credit: D. Kelletat)

Fig. 12.26 (continued)

c

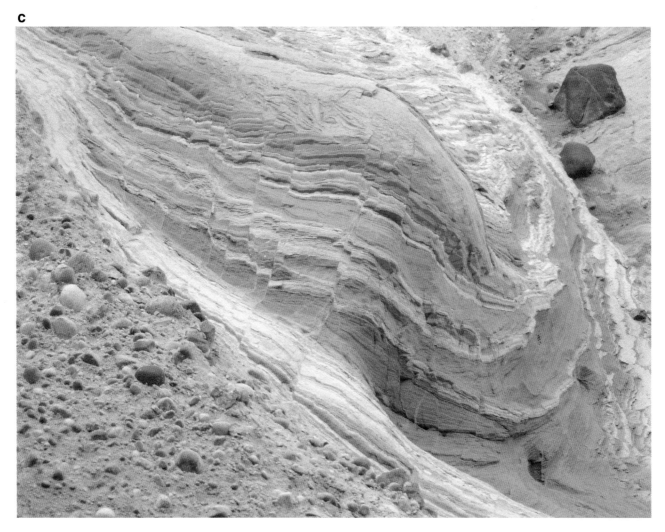

Fig. 12.26 (continued)

a

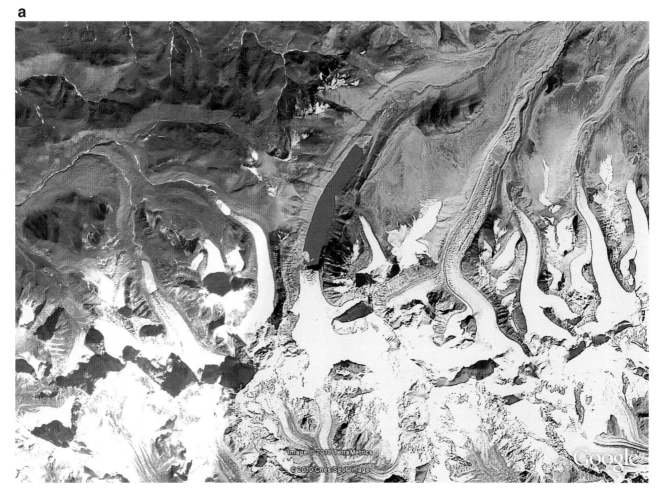

Fig. 12.27 (**a**) Lakes, as well as lateral and terminal moraines, in front of glaciers in southernmost Tibet document a young glacier retreat at *28°15′N* and *86°17′E*. Scene is about 35 km wide. (**b**) Well preserved lateral and terminal moraines near *37°50′N* and *119°05′W* along the east side of California's Sierra Nevada recording the maximum extent of glaciers during the last glacial maximum advance and some late glacial re-advances. Scene is 20 km wide. (**c, d**) Terminal and lateral moraines from receding glaciers in the Himalayas at about *27°55′N* and *87°45′E* in (**c**), and *28°23′N* and *86°23′E* in (**d**). Lakes developed behind the wall of the terminal moraine by meltwater. (Image credit: ©Google earth 2012)

b

c

Fig. 12.27 (continued)

d

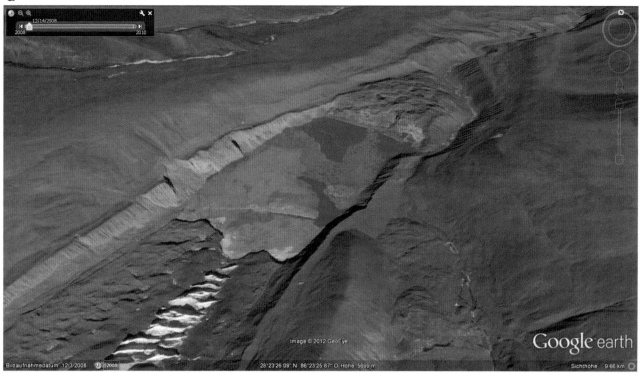

Fig. 12.27 (continued)

Irish language; Fig. 12.28) are usually more narrow and longer than roches moutonnées.

Glaciofluvial processes involve subglacial meltwater discharge at the base of the ice that are capable of transporting large amounts of well-rounded pebbles and rock debris in meltwater tunnels. Where these deposits accumulate during deglaciation and their form is not modified by subsequent glacial advances, very long ridges of pebbles may develop, sometimes even showing bifurcation patterns. These landforms may be several meters high, up to some decameters wide and kilometers long (Fig. 12.29). They are called "*osers*" (from Swedish) or "*eskers*" (from Irish). In depressions between a decaying glacier and the valley slopes, accumulations of rather chaotic debris may occur (Fig. 12.30). When a glacier recedes, terrace-like features may be left, forming "*kames*". Finally, slightly sloping sediment fans of mostly sandy meltwater debris in front of the terminal moraine are called "*sander*", "*sandr*" or "*sandur*" (from Icelandic). In areas of former glaciation, the entire series of these glacier-related landforms and deposits may be preserved. This is particularly common for areas affected by the glaciers of the last glacial period that are in landscapes with a low topography.

12 Glacier Ice and Its Domain

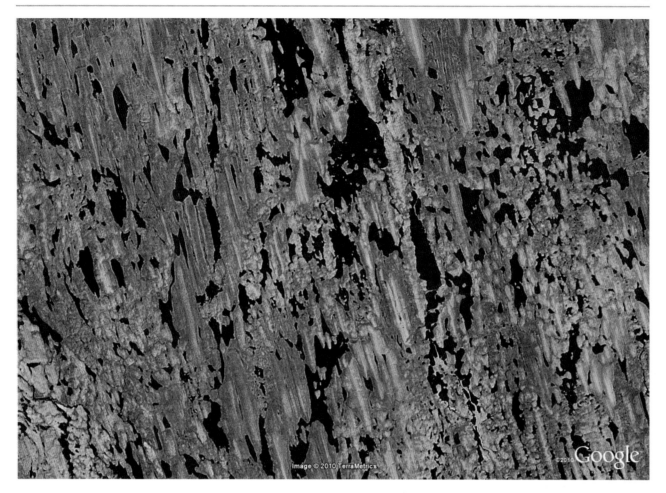

Fig. 12.28 Fields of drumlins in western Canada in a 30 km wide scene at *57°10′N* and *67°45′W* (Image credit: ©Google earth 2012)

Fig. 12.29 Elongated ridges (osers or eskers) of loose glacial and meltwater deposits in Canada at *52°15′N* and *99°19′W* in an area at least 70 km wide (Image credit: ©Google earth 2012)

Fig. 12.30 Well developed lateral moraine and fluvial accumulation of debris in the depression between moraine and valley slope. This may lead to the evolution of kames deposits. The scene is from the New Zealand Alps on South Island (Image credit: ©Google earth 2012)

Further Readings

Anderson B (2003) The Response of Franz Josef Glacier to climate change. University of Canterbury, Christchurch

Anderson D (2004) Glacial and periglacial environments. Hodder Arnold, London

Andreassen LM, Elvehøy H, Kjøllmoen B, Engeset RV, Haakensen N (2005) Glacier mass balance and length variations in Norway. Ann Glaciol 42:317–325

Bamber JL, Layberry RL, Gogineni SP (2001) A new ice thickness and bed data set for the Greenland ice sheet, 1. Measurement, data reduction, and errors. J Geophys Res 106:33733–33780

Benn D, Evans D (1998) Glaciers and glaciation. Hodder Arnold, London

Benn DI, Warren CR, Mottram RH (2007) Calving processes and the dynamics of calving glaciers. Earth Sci Rev 82(3–4):143–179

Bennett MR, Glasser NF (1996) Glacial geology: ice sheets and landforms. Wiley, Oxford

Bingham RG (2010) Glacial geomorphology: towards a convergence of glaciology and geomorphology. Prog Phys Geogr 34(3):327–355

Brodzikowski K, van Loon AJ (1987) A systematic classification of glacial and periglacial environments, facies and deposits. Earth Sci Rev 24(5):297–381

Chinn TJH (1988) Glaciers of New Zealand. In: Satellite image atlas of glaciers of the world. U.S. Geological Survey, Professional Paper 1386. Reston, Virginia (USA)

Clark CD, Hughes ALC, Greenwood SL et al (2009) Size and shape characteristics of drumlins, derived from a large sample, and associated scaling laws. Quat Sci Rev 28:677

Deynoux M, Miller MGJ, Domack EW et al (eds) (2004) Earth's glacial record. University Press, Cambridge

Haeberli W (2007) Changing views on changing glaciers. In: Orlove B, Wiegandt E, Luckman B (eds) The darkening peaks: glacial retreat in scientific and social context. UC Press, Berkeley, pp 23–32

Haeberli W, Holzhauser H (2003) Alpine glacier mass changes during the past two millennia. PAGES News 11(1):13–15

Hagen JO, Kohler J, Melvold K, Winther JG (2003) Glaciers in Svalbard: mass balance, runoff and fresh water flux. Polar Res 22(2):145–159

Johnson MD, Schomacker A, Benediktsson IO et al (2010) Active drumlin field revealed at the margin of Mulajokull, Iceland: a surge-type glacier. Geology 38:943

Martini IP, Brookfield ME, Sadura S (2001) Principles of glacial geomorphology and geology. Prentice Hall, Upper Saddle River

Menzies J (ed) (1995) Modern glacial environments: processes, dynamics and sediments. Butterworth-Heinemann Ltd., Oxford

Naruse R (2006) The response of glaciers in South America to environmental change. In: Knight PG (ed) Glacier science and environmental change. Blackwell, Oxford

NSIDC (2008) World glacier inventory. Data Set Documentation. World Glacier Monitoring Service and National Snow and Ice Data Center/World Data Center for Glaciology. Boulder. http://nsidc.org/data/docs/noaa/g01130_glacier_inventory/. Accessed 18 May 2014

Oerlemans J (2001) Glaciers and climate change. A.A. Balkema Publishers, Lisse/Abingdon/Exton/Tokyo

Paterson WSB (1994) The physics of glaciers, 3rd edn. Pergamon Press, Oxford

Patterson CJ, Hooke RL (1995) Physical environment of drumlin formation. J Glaciol 41:30–38

Rignot E, Kanagaratnam P (2006) Changes in the velocity structure of the Greenland ice sheet. Science 311(5763):986–990

Shaw J (2002) The meltwater hypothesis for subglacial bedforms. Quat Int 90:5–22

Shaw J, Kvill D (1984) A glaciofluvial origin for drumlins of the Livingstone Lake area, Saskatchewan. Can J Earth Sci 21:1442

Solomina O, Haeberli W, Kull C, Wiles G (2008) Historical and Holocene glacier-climate variations: general concepts and overview. Global Planet Change 60(1–2):1–9

Spagnolo M, Clark CD, Hughes ALC, Dunlop P, Stokes CR (2010) The planar shape of drumlins. Sediment Geol 232:119–129

Sugden DE, John BS (1976) Glaciers and landscapes. A geomorphological approach. Wiley, Oxford

13 Frost and Permafrost as Morphological Agents (or: the Periglacial Domain)

Abstract

Frost is not only an important agent to destroy rocks by frost shattering, but as permanent frost (permafrost) in the ground (either in sediment, soil or even hard rock) it also contributes to distinctive patterns in the landscape (called "patterned ground"). Some are only centimeters wide, while others can be seen for many square kilometers. The extension of modern permafrost on the northern hemisphere alone is over 12 million km². It also appears on the upper shelf under the oceans, left behind from the glacial periods where it was exposed due to the lower sea level. According to the thickness, continuity and extension, we separate permafrost landscapes into those with continuous, discontinuous and sporadic permafrost. In modern times with rising temperatures, *thermokarst* as an increasingly important phenomenon may affect permafrost areas.

We can classify all frost-related forms on the surface into those made by strong winter frosts (where freezing is the active process), or those made by permafrost (where thawing of the so-called "active layer" in summer is the active process). Another differentiation is possible between forms on slopes such as ice-related solifluction or gelifluction (i.e., the slow flowing of water-saturated soil material and debris above a frozen basal layer) with typical solifluction terraces and lobes, and features made against gravity by the extension of freezing water in the ground. Small frost hills, larger *palsa* and even large *pingos* (up to 30 m high) are examples for the latter forms. We may also distinguish between processes and forms with or without sorting of particles; typical substrate sorting in periglacial areas produces rather regular forms with fine particles in the center and coarser particles around (on a plain as networks or polygons, or as stripes of fine and coarse material on slopes). *Ice wedges* in cracks of deeply frozen permafrost may form extended polygonal patterns in flat landscapes of northern Siberia, Alaska or Canada. In high mountains with low winter temperatures, tongues of debris may move from debris cones showing glacier-like forms ("rock glaciers"). Lastly the decay of permafrost, promoted by the ongoing climate change and global warming, produces countless ponds in a permafrost landscape in sizes ranging from many kilometers in diameter to small-scale forms of only a meter or so. Again we use mainly terrestrial photographs to demonstrate the smaller forms, but some of the patterned grounds in permafrost regions also can be detected by satellites.

Periglacial landforms from the cold and frozen grounds of high latitudes and altitudes are widely unknown to the public because they often are located in remote regions without any infrastructure for access. These regions are extremely cold in winter, but due to higher temperatures in summer the surface is wet and swampy because of thawing processes on the ground ice. As permafrost layers are impermeable, water cannot percolate into the ground; it only disappears by evaporation or, if inclination of the surface is present, flows into rivers. Nevertheless, permafrost-related processes are important in geomorphology because they were actively shaping the Earth during all the glacial periods, and they are still active in present day temperate latitudes. Due to the characteristic landforms and sedimentation patterns they produce, geoscientists can easily find the imprints of periglacial processes in the landscape.

Periglacial forms may be generated in the periphery of ice (i.e., very cold areas), but without direct glacier contact. The most extensive areas of present day permafrost are in accordance with today's cold climates in northern Alaska, Canada or Siberia. Thus, the periglacial zones of the Earth

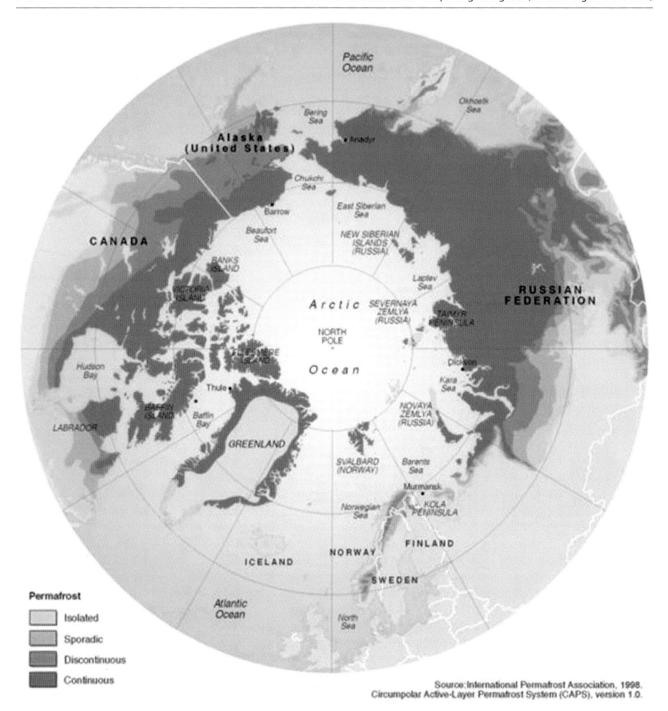

Fig. 13.1 Extension of permafrost types in the arctic and sub-arctic environments (Image credit: Philippe Rekacewicz, 2005, UNEP/GRID-Arendal Maps and Graphics Library)

are very extensive and cover more than half of Alaska, Canada and Russia (Fig. 13.1). Permafrost occurs in eastern Russia in latitudes south of that of New York in the USA (Fig. 13.2). During the glacial periods of the Pleistocene, periglacial processes occurred as far south as the Mediterranean and southern USA as well as down to the foothills of mountains ranges, because temperatures have been 10–12 °C lower than today on average. Permanently frozen ground, up to several 100 m deep in rocks, may also be relictic from the glacial periods in regions without glaciation (because of dry climates with too little snow to form glaciers). Permafrost also occurs in high mountains, from central Asia and the Himalayas to the Andes or the European Alps. We discriminate between continuous permafrost, discontinuous permafrost and sporadic permafrost (Figs. 13.1 and 13.2).

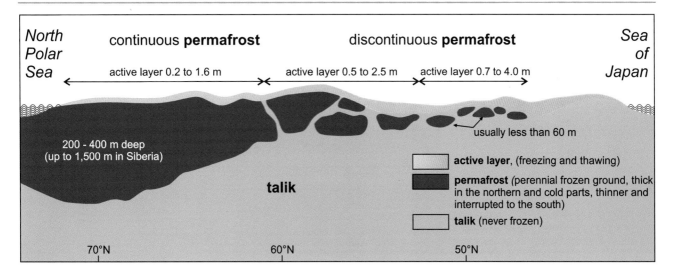

Fig. 13.2 A north–south section across the permafrost landscape in eastern Siberia (Russia). In the north, permafrost is several hundred meters deep and continuous. Towards lower latitudes, it is becomes thinner and shows gaps, whereas at even warmer latitudes it is separated into isolated patches and is only a few meters deep. As climate change affects permafrost from the surface, the thawing processes may accelerate permafrost decay in all latitudes (Image credit: A. Hager)

In high latitudes and altitudes, the change between freezing and thawing (even in permafrost, where thawing at the surface is restricted to the "active layer" during summer) causes numerous small- and medium-sized forms. Many of them follow the slopes (like solifluction lobes; see Figs. 13.3, 13.4, 13.5, and 13.6), while others form against gravity by ice extension (during freezing) and uplift of debris, soil material and vegetation (Figs. 13.7, 13.8, 13.9, 13.10, and 13.11). We can sum up all these forms into the category of "patterned ground", because they exhibit forms with a particular organization pattern, like pictures of artificial surfaces.

Distinctive patterns due to a sorting process that differentiate between fine and coarse particles may exhibit different sizes, from decimeters to several meters (Figs. 13.12 and 13.13). Firstly, larger rocks are lifted by recurring frost-and-thaw-cycles to the surface; then, the surface forms earthy bumps just centimeters high, appearing as earthy patches in the ground. The lifted coarse particles are shifted from the center of the bumps to the peripheries, mainly by needle ice (because of the slopes of the earthy bumps). Here, the larger rocks accumulate and form a ring or a polygon. Large networks of these rings and polygons characterize wide areas with flat surfaces in the periglacial zone. The same process of particle sorting is responsible for the generation of stone stripes on slopes (Figs. 13.14a, b).

We know that typically a lot of materials shrink when they cool down. This also is true for frozen ground and ice. With frost of about −20 to −40 °C in winter, cracks of several centimeters width are opened by contraction in the upper parts of the frozen ground, closing at depths of a few meters. In springtime, melt water drops in these open cracks and immediately freezes along the walls, widening the fissures due to expansion and pressure during the transformation into ice. Within a few years open wedges are formed, centered by pure ice or ice mixed with fine particles, and polygonal networks of ice wedges (with polygons of several meters to decameters across) characterize wide and flat periglacial landscapes (Figs. 13.15, 13.16, and 13.17).

Where swamps or bogs freeze deeply in winter, the peat is under pressure from the extending ice centers and deforms into stripes or strings. The so-called "string bogs" are further documentation for the permafrost processes (Fig. 13.18).

As permafrost is defined as the frost surviving the summer periods, each glacier is a permafrost body; this is also true for "rock glaciers", which are streams of slow moving coarse debris with interstitial ice, mostly derived from snow meltwater (Fig. 13.19a–d). Rock glaciers are nourished from frost weathering e.g. along old cirque walls. A debris cone at the foot of these walls – if it occurs at permafrost levels – will start to move in the form of a thick tongue of boulders. Similar to real glaciers, patterns of their movement or deformation are produced on their surface; hence, these features have the perfect name "rock glaciers". If weathering rates and rockfall is reduced, the nourishment of the rock glaciers also will be reduced and their movement may stop. The age of all these rock glaciers is of post-glacial times, because during the last period of glaciation (terminating ~12,000 years before present), the cirques have been filled by glacier ice.

So far we have presented images from permafrost-related forms and we discussed the most important processes. Since we are living in a period of rising temperatures, particularly in the high latitudes of the northern hemisphere, accelerated melting of former permafrost features is observed today. We can summarize these features under

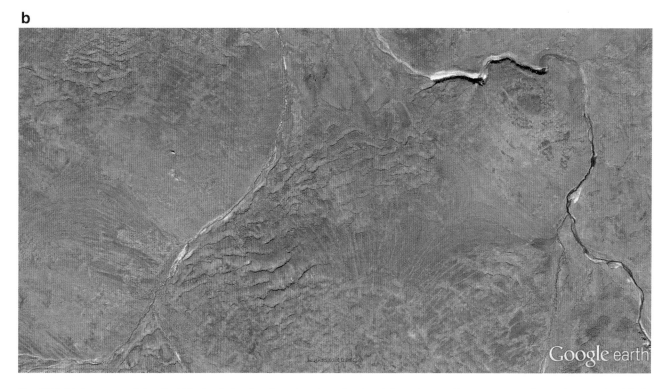

Fig. 13.3 (**a**) Loose material and soil is slowly flowing downslope in northernmost Norway and solifluction lobes as a kind of "patterned ground" occur. Movement may be the result of thawing (with water saturation and low friction following gravitational forces), or of a dislocation downwards by extension of the debris and soil material during freezing (Image credit: D. Kelletat). (**b–d**) Broad solifluction lobes characterize numerous slopes in western Alaska (65°22′23.51″N, 163°31′0.02″W and 65°22′23.05″N, 163°11′47.47″W). Stone stripes and stone polygons as seen in (**c**) are typical of periglacial landscapes as well (see also Figs. 13.12, 13.13, and 13.14) (**d**) is a detail of the image in (**c**) (Image credit ©Google earth 2012)

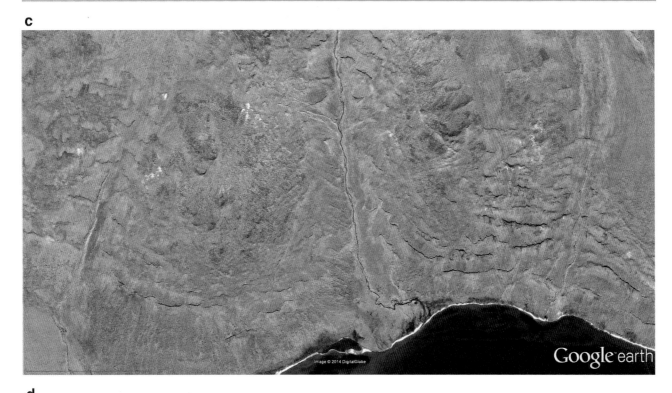

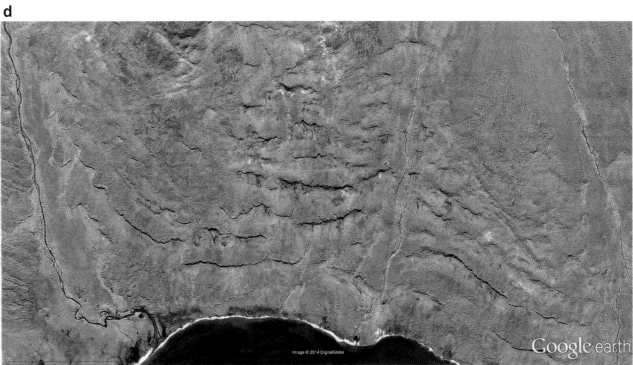

Fig. 13.3 (continued)

Fig. 13.4 Frost shattered debris from the last ice age has been slowly incorporated in the creeping process of solifluction showing a bending of rock or soil strata. Today this process may be inactive at this location because of absence of in frost regime. Example is from southern Ireland (Image credit: D. Kelletat)

Fig. 13.5 Solifluction is the slow flowing of unconsolidated debris and soil material downslope. It denudates and smoothens mountains and hills, forming an accumulation of rock material at the base of the slopes. Scene is 15 km wide from western Alaska (USA) at *66°02′N* and *159°18′W* (Image credit: ©Google earth 2012)

Fig. 13.6 Periglacial landscape with gelifluction/solifluction south of James Bay, Canada, at *51°25'N* and *78°43'W* in a 16 km wide scene. Although it is a slow process with movements of centimeters to a few decimeters per year, solifluction leaves distinctively smooth landscapes and is a very effective denudation process in cold climate environments (Image credit: ©Google earth 2012)

Fig. 13.7 Ice needles have lifted some dirt and debris. They grow upwards from soil moisture in cold nights, and during the day the debris may creep or even flow as the ice melts due to water saturation (Image credit: D. Kelletat)

Fig. 13.8 (a) Earthy hummocks about 0.5 m high in the Austrian Alps at 2,400 m above sea level, formed by the uplifting of rock and soil material during the freezing of pore waters and the expansion of water. These features (called "*thufur*" in Iceland) may form within 2–3 winter periods. (**b**) Thufur landscape in central Iceland. Frost hills are up to 0.8 m high and 1.5–2.5 m across (Image credit: D. Kelletat)

b

Fig. 13.8 (continued)

Fig. 13.9 (**a**) Larger forms of frost hills, often concentrated in peat bogs, may reach heights of 2–3 m and diameters of tens of meters. They are either plateau-like with a flat surface and tens of meters across, or dome-shaped. They are called "*palsas*" from a Swedish term. This example is from the northeastern-most parts of Norway at *70°N*. Under the cover of peat, which isolates the frozen parts from thawing in summer, is a pure ice body. Scale is 1 m, see measure tape along the wall (Image credit: D. Kelletat). (**b**) Earth hummocks (most probably thufurs or palsas) have formed in the center of this image from northwestern Alaska (*65°22′52.26″N* and *163°24′25.70″W*). Image is ~1.3 km wide, top of image is south (Image credit: ©Google earth 2012)

13 Frost and Permafrost as Morphological Agents (or: the Periglacial Domain)

b

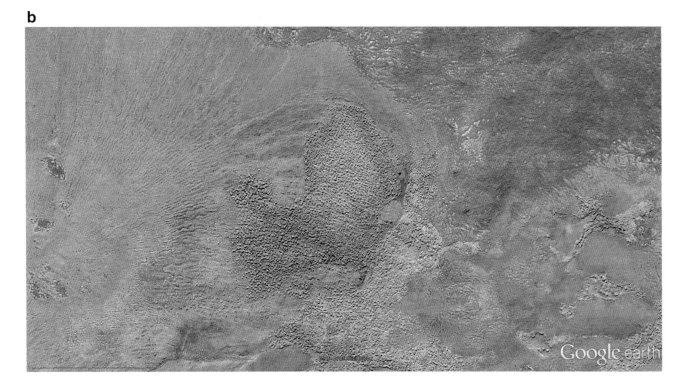

Fig. 13.9 (continued)

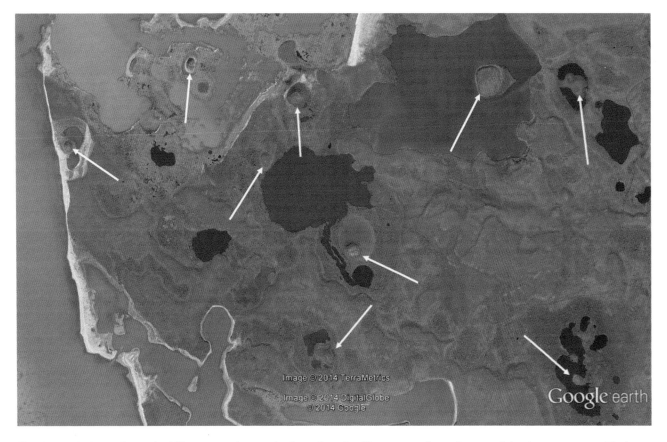

Fig. 13.10 The largest frost or ice hills grown against gravity (in northernmost Alaska, Siberia, Greenland and Spitsbergen) are termed "*pingos*" (from an Inuit term in Greenland). These are pure ice hills up to more than 20 m high in river beds, swamps or bogs, and grow by hygroscopy of ice to attract more and more water from the immediate environment. By this process, a depression around the pingos may occur. The image shows 9 pingos on less than 10 km^2 in the Mackenzie River delta of northwest Canada; arrows indicate location of pingos (Image credit: ©Google earth 2012)

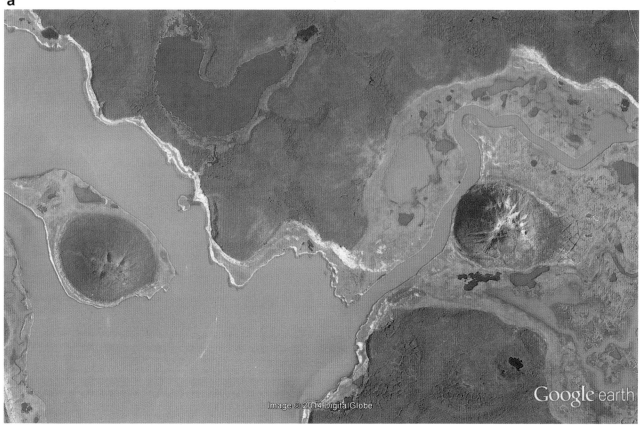

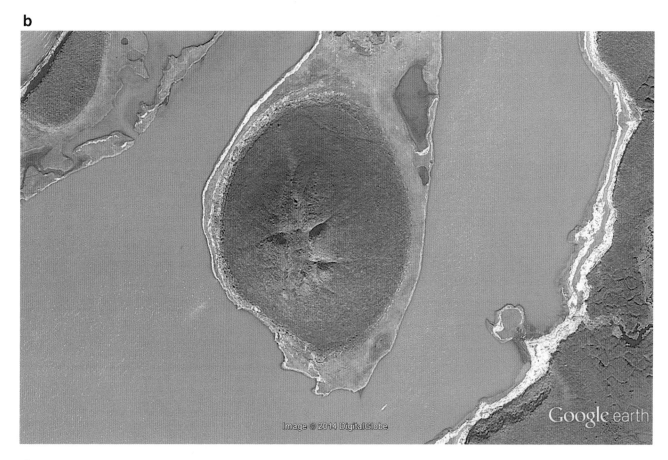

Fig. 13.11 Close-up views of pingos from the Mackenzie River delta, northwest Canada. The "twins" (**a–c**) are at *69°24′N* and *133°05′W*, and the relic pingo (**d**) is at *70°10′51.20″N* and *129°40′52.75″W*. The diameter of the pingos at this latitude may reach 100 to >300 m (Image credit: ©Google earth 2012)

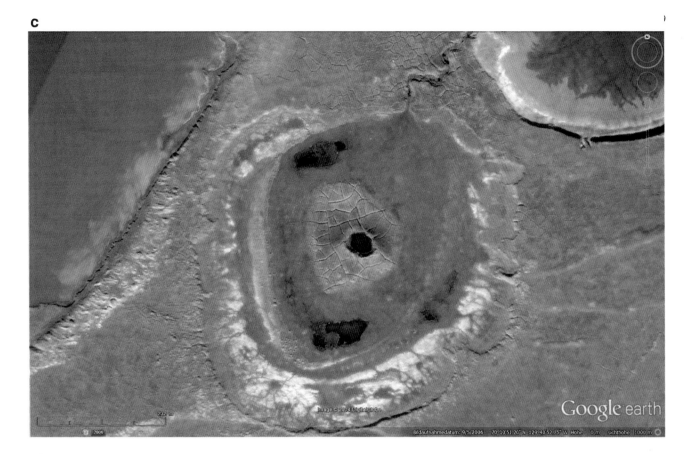

Fig. 13.11 (continued)

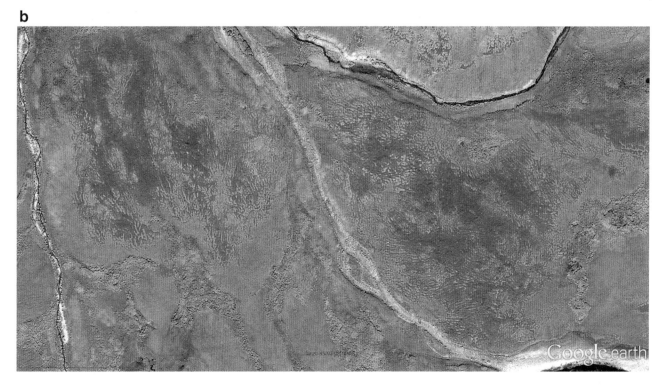

Fig. 13.12 (a) Stone polygons formed by multiple freezing and thawing cycles which at first move the coarse particles to the surface, where they are further separated from the fine ones forming networks or polygons (on flat ground). The coarse fragments form the outer frames. This example is from the Scottish Highlands at 1,200 m above sea level on weathered granitic rock. The diameter of each earthy center is only about 20 cm (Image credit: D. Kelletat). (**b, c**) Patterned ground due to stone polygons captured with Google Earth in western Alaska (65°23′29.94″N, 163°29′6.20″W and 65°22′3.25″N 163°23′24.18″W). Stone stripes develop where the surface starts to slope, as visible in the upper left of the image in (**b**) (Image credit: ©Google earth 2012)

c

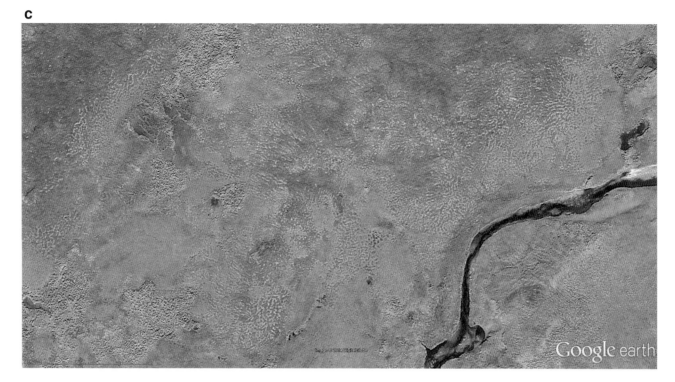

Fig. 13.12 (continued)

Fig. 13.13 Patterned ground with fine centers and coarse frames may reach individual diameters up to more than 3 m as in this example from the high fjell areas of southern Norway, close to 2,000 m above sea level. Generally, the dimensions of single stone rings or stone polygons are larger in deep winter frost and particularly in permafrost regions, and small only if short frost periods occur (Image credit: D. Kelletat)

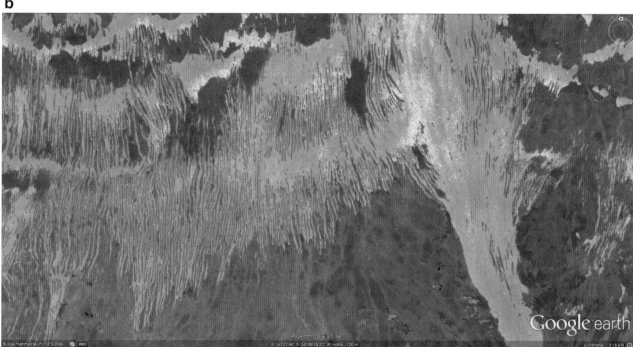

Fig. 13.14 (a) Networks on slopes are transformed into stripes by gravitational force, again showing sorting of fine and coarse material. Image from the Scottish Highlands with lengths of the coarse stripes of 20–30 m (Image credit: D. Kelletat). (b) Stone stripes on hill slopes of the Falkland Islands in the southern Atlantic Ocean at about *51°43'S* and *58°10'W*. Stripes are up to 1 km long with distances up to 30 m (Image credit: ©Google earth 2012)

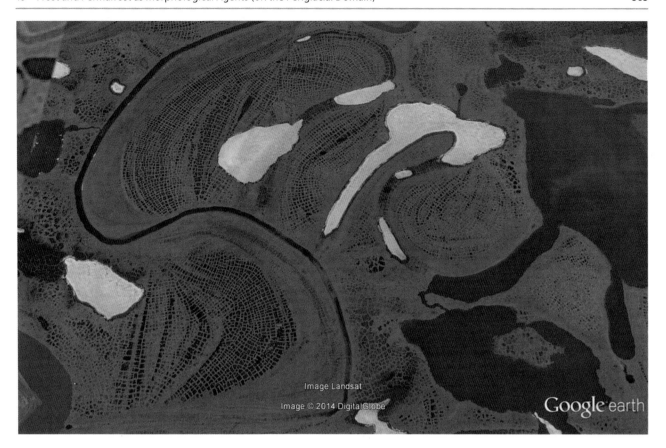

Fig. 13.15 With low temperatures (−25 to −40 °C) in winter, the frozen ground shrinks repeatedly and cracks open up in networks, which will be filled by meltwater in the warmer times of the year. The water travels deeper and opens the cracks further in each thawing/freezing cycle. The cracks show more or less pure ice in the form of ice wedges and this kind of patterned ground is called "ice wedge polygons". The site is located in northern Alaska at *70°50′N* and *155°30′W* and the scene is about 5 km wide (Image credit: ©Google earth 2012)

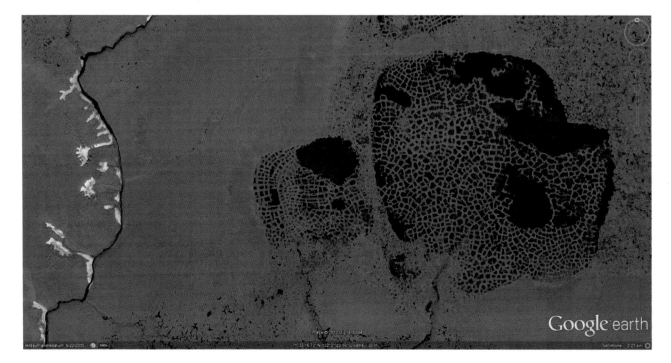

Fig. 13.16 Irregular network of patterned ground by ice wedge polygons in a shallow depression of northern Russia at *71°35′N* and *132°32′E*. Scene is 2.3 km wide (Image credit: ©Google earth 2012)

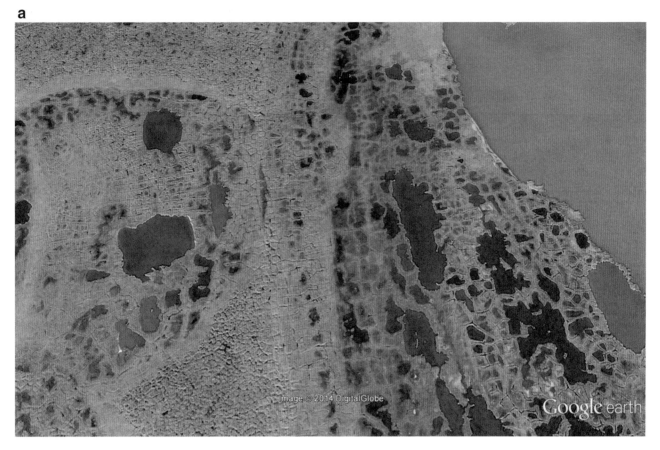

Fig. 13.17 (a) Ice wedge polygons in the permafrost of northern Alaska. Scene is 1 km wide at *70°55'27.57"N* and *155°19'35.19"W*. The overall pattern evidently is adapted by the sediment character and/or slight topography from the nearby river. Single polygons have diameters up to 40 m. (b) A 1.5 km wide scene at *70°22'22.48"N* and *151°01'41.51"W*. The pattern is irregular with a diameter of polygons of 15–35 m. We can identify dark lines within the elevated rims of the polygons which are melting ice wedges, whereas the elevation is the result of the pressure by actively freezing vertical fillings of fissures. As along each fissure a double low rim is formed, the polygons form a network of low-centered features that are filled by water in summer and frozen over winter. Freezing of the ponds also will exert pressure towards their outer rims and elevate the network lines, which may be under the influence of pressure from both sides (Image credit: ©Google earth 2012)

b

Fig. 13.17 (continued)

a

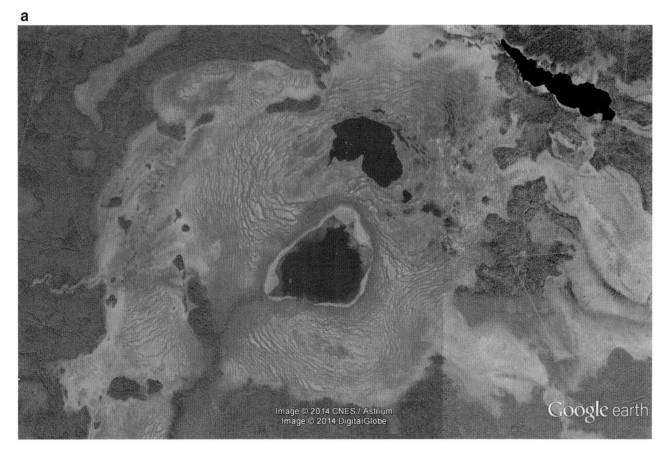

Fig. 13.18 "String bogs" (deformation of peat layers by pressure during freezing cycles) in western Russia: (**a**) 4.5 km scene at *64°21′N and 42°08′E*, (**b**) 2 km wide scene at *64°41′N and 42°18′E*. Both patterns are the result of low sloping of the surface, and the deformation of the strings in the peat bogs develop as a kind of contour lines just a few centimeters different in altitude (Image credit: ©Google earth 2012)

b

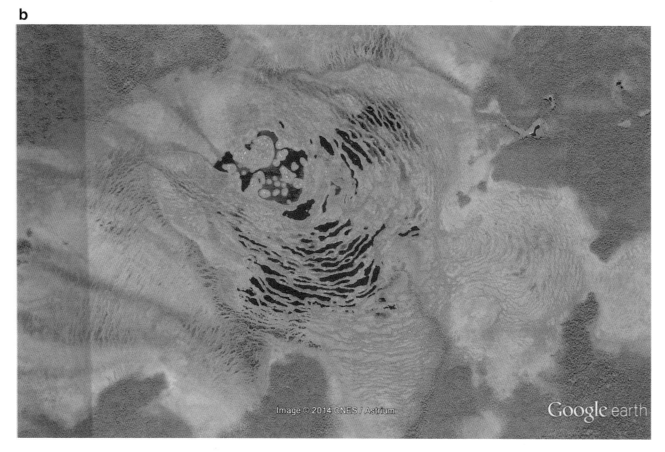

Fig. 13.18 (continued)

a

Fig. 13.19 (**a–d**) A special kind of debris movement called "rock glaciers", made of masses of coarse frost debris with interstitial ice from meltwater. The ice fillings of the wide pores between boulders allow the rock glaciers to overcome friction, resulting in a slow movement of 1–5 m per year. This movement is shown by the string-like surface structures on the debris mass. If actively moving, the fronts are normally steep and, except of some lichen, no vegetation is present. Examples are from the Sierra Nevada of eastern California (USA; *37°53′45″N, 119°12′09″W*), the Hindukush in Afghanistan (*36°39′50″N, 72°02′24″E*) southern Chile (*30°09′15″S, 69°54′40″W*), and the Swiss Alps (*46°30′21″ N, 09°55′44″E*) (Image credit: ©Google earth 2012)

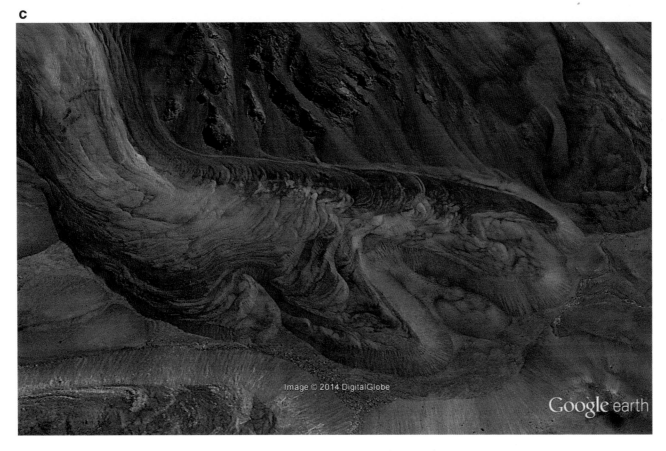

Fig. 13.19 (continued)

d

Fig. 13.19 (continued)

the term of "*thermokarst*", which refers to a suite of landforms that is created from the shrinking or degradation of the permafrost (Figs. 13.20, 13.21, and 13.22). A special form of thermokarst results from the "thermo-abrasion" of shorelines where cliff retreat in grounds composed of frozen debris takes place, and where the permafrost is mechanically removed by wave action. This process of coastal erosion endangers coastal settlements and other infrastructure. It seems to have accelerated within the past years, and rates of 20–50 m of land loss per year have been reported. This process also connects thermokarst lakes to the Arctic Ocean.

In latitudes with dominant wind direction, thermokarst lakes are indicated by elongated forms and evidently are enlarged around their shorelines creating a "cat-eye" appearance (Fig. 13.22). Decaying permafrost also poses an increased hazard factor in high mountain regions. Thawing within the rock weakens its stability and may lead to increased size and frequency of rockfalls and rockslides, endangering settlements in the valleys below.

a

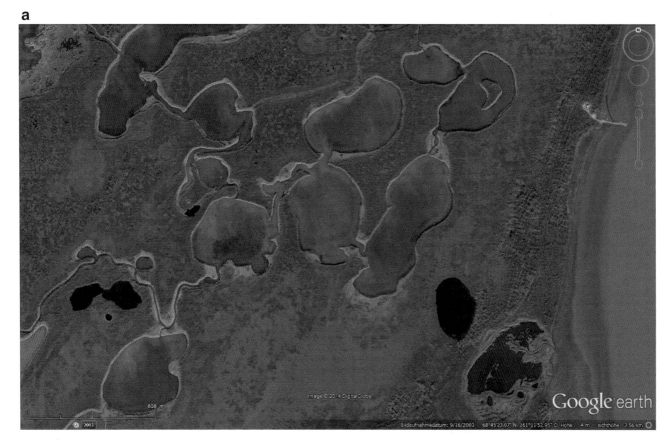

Fig. 13.20 Two examples of recent thermokarst, which means thawing of permafrost. (**a**) Patches of the former permanently frozen ground are now collapsed by thawing, and new lakes with irregular but sharp contours (ice and debris cliffs) occur. Scene is 4 km wide, at *68°45'N, 161°11'E* (**b**) Destruction of an older depression. Scene is 5.2 km wide at *69°38'N, 162°27"E*. Both examples are from northern Siberia (Image credit: ©Google earth 2012)

b

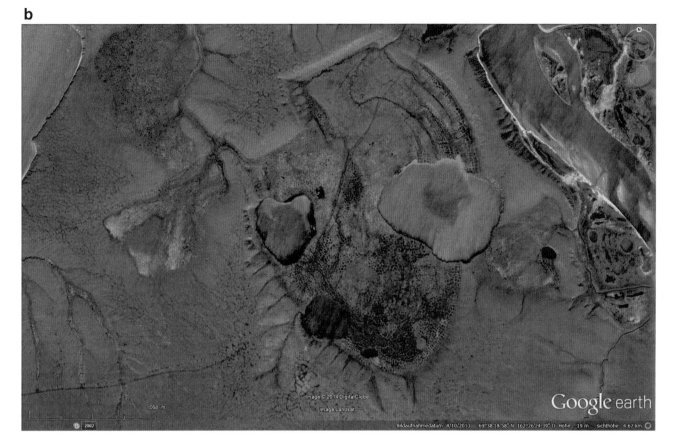

Fig. 13.20 (continued)

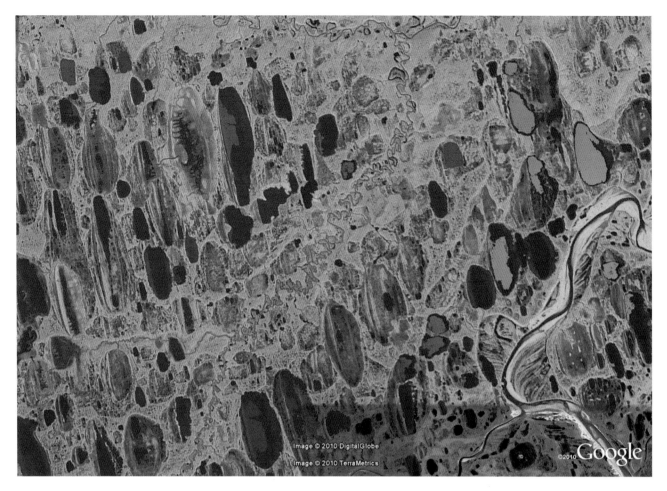

Fig. 13.21 Elongated lakes in a permafrost region, formed by constant wind direction during the thawing period, show ongoing permafrost decay in Alaska, northern Canada and Siberia (Image credit: ©Google earth 2012)

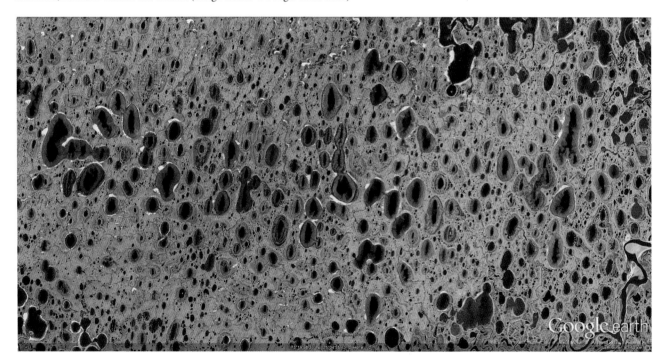

Fig. 13.22 "Cat-eye" thermokarst lakes in the Lena Delta of northern Siberia, Russia, at around *73°19′N, 125°35′E*. Width of scene is 32 km. The effect is a result of the sideward extension of the lakes in all directions due to higher water temperatures and longer melting phases (Image credit: ©Google earth 2012)

Further Readings

Anderson D (2004) Glacial and periglacial environments. Hodder Arnold, London

Barsch D (1996) Rockglaciers. Indicators for the present and former geoecology in high mountain environments. Springer, Berlin

Boelhouwers H, Holness S, Sumner P (2003) The maritime subantarctic: a distinct periglacial environment. Geomorphology 52(1–2):39–55

Brodzikowski K, van Loon AJ (1987) A systematic classification of glacial and periglacial environments, facies and deposits. Earth Sci Rev 24(5):297–381

Clark DH, Steig EJ et al (1998) Genetic variability of rock glaciers. Geogr Ann 80(3–4):175–182

Davis N (2001) Permafrost: a guide to frozen ground in transition. University of Alaska Press, Fairbanks

French HM, Guglielmin M (1999) Observations on the ice-marginal, periglacial geomorphology of Terra Nova Bay, Northern Victoria Land, Antarctica. Permafr Periglac Process 10(4):331–347

Giardino JR, Shroder JF et al (1987) Rock glaciers. Allen & Unwin, London

Haeberli W (1992) Construction, environmental problems and natural hazards in periglacial mountain belts. Permafr Periglac Process 3:111–124

Humlum O (1996) Origin of rock glaciers: observations from Mellemfjord, Disko Island, Central West Greenland. Permafr Periglac Process 7:361–380

Humlum O (1999) The climatic significance of rock glaciers. Permafr Periglac Process 9(4):375–395

Humlum O (2000) The geomorphic significance of rock glaciers: estimates of rock glacier debris volumes and headwall recession rates in W Greenland. Geomorphology 35:41–67

Humlum O, Instanes A, Sollid JL (2003) Permafrost in Svalbard: a review of research history, climatic background and engineering challenges. Polar Res 22(2):191–215

Kessler MA, Werner BT (2003) Self-organization of sorted patterned ground. Science 299(5605):380

Konrad SK, Humphrey NF et al (1999) Rock glacier dynamics and paleoclimatic implications. Geology 27(9):1131–1134

Matsuoka N (2001) Solifluction rates, processes and landforms: a global review. Earth Sci Rev 55(1–2):107–134

Matsuoka N, Ikeda A, Date T (2005) Morphometric analysis of solifluction lobes and rock glaciers in the Swiss Alps. Permafr Periglac Process 16(1):99–113

Van Everdingen RT (ed) (2005) Multi-language glossary of permafrost and related ground-ice terms. National Snow and Ice Data Center/World Data Center for Glaciology, Boulder

Washburn LA (1980) Geocryology: a survey of periglacial processes and environments. Wiley, New York

Whalley WB, Azizi F (2003) Rock glaciers and protalus landforms: analogous forms and ice sources on earth and mars. J Geophys Res 108:1–17

Williams PJ, Smith MW (1989) The frozen earth. Cambridge University Press, New York

Part IV

Epilogue

14. Transformation of the Earth's Surface by Man (Anthropogenic Forms)

Abstract

Since the beginning of the times of industrialization man has transformed large parts of the global surface by technical installations such as dams, quarries, and open pit mining. However, a direct geomorphologic transformation of the slopes has also taken place since thousands of years and in very large areas for the cultivation of crop (and wet rice farming) in the form of artificial terraces, which cover several million square kilometers today. Besides the direct transformation of existing forms, indirect impacts by deforestation and overgrazing also influence runoff, weathering and all the different transport processes. Due to over-pumping of groundwater and overgrazing as well as the rising population and the end of nomadism, "*desertification*" (man-made deserts) is rapidly becoming one of the greatest threats of the future.

Since thousands of years man has been active in forming parts of the Earth's surface. The most important direct influence certainly is the transformation of slopes into a staircase of terraces to maximize areas for cultivation and irrigation (Figs. 14.1, 14.2, and 14.3), which also deeply influences sediment transport into rivers. This measure started with the first crop cultivation more than 8,000 years ago in the Near East and is common in rice-growing landscapes all around the world. It is still continuing today, but now with much wider consequences, because machines form much larger terraces in a very short time. Other direct impacts on the relief are mining areas from the surface up to more than 500 m deep in open pits that are kilometers in length, think of the mining for gold in South Africa, for copper, silver and gold in Chile, for iron and gold in Australia, and for hydrocarbons in many other countries (Figs. 14.4, 14.5, 14.6, 14.7, 14.8, 14.9, 14.10, and 14.11). This open pit mining also influences groundwater levels and the hydrology of large areas. Combined with these mining activities, a huge amount of rock has to be deposited on nearby sites, forming a new hilly landscape. Even deep mining with shafts may transform the landscape, because after decades the tunnels and galleries under the surface break down and as a consequence the surface may subside, as in parts of Germany for more than 10 m within less than 100 years. Open pit mining in a humid climate requires the groundwater level to be reduced significantly, but after the end of mining groundwater will rise again, and the now lowered sites are strongly affected by flooding hazards.

More indirect impacts on the natural geomorphology are sand and pebble mining in rivers where the removal of bed load and delta deposits interrupts or at least disturb the natural cycle of erosion and deposition. The same is the case if rivers are fixed by dikes or coastlines are changed by land reclamation. Thousands of quarries are opened in every country to collect different kinds of rock for construction purposes. These roads cut natural slopes and interrupt natural debris dislocation and distribution (Fig. 14.12). The correcting and shortening of river courses for navigation purposes (Fig. 14.13) may accelerate flow and sediment transport but also prevents large areas from natural flooding. Large and small reservoirs (for drinking water, irrigation, energy) become intermediate places of deposition in a river's course, where formerly all sediments would have been transported further downstream (Fig. 14.14). Literally millions of square kilometers have been changed by deforestation which accelerates sediment transport, and this is an ongoing process across many agro-industries. A rising population on our globe and the huge appetite for space for agriculture and energy (e.g., deforestation, mining, reservoir construction,

Fig. 14.1 In many countries of the tropics, wet rice culture is extremely extensive, and millions of terraces have been built for irrigation. This is an example from Bali, Indonesia (Image credit: D. Kelletat)

Fig. 14.2 Agricultural terraces in eastern Java at about *7°09′S* and *109°31′E* (Image credit: ©Google earth 2012)

Fig. 14.3 In the northwestern parts of China, the loess landscape is transformed into terraces for crops and vegetables (*38°52′N, 111°41′E*) (Image credit: ©Google earth 2012)

Fig. 14.4 Part of Coober Pedy (Australia) at *29°01′S* and *134°44′E*, in a 1 km wide scene. Dozens of shallow opal mines (>250,000 mine shafts) have been dug out so far. It is the world's largest opal mining area (Image credit: ©Google earth 2012)

Fig. 14.5 Sand mining for construction purpose in southern Jordan at *31°01′N* and *36°04′E*. Image shows a 4.2 km wide section of the desert landscape (Image credit: ©Google earth 2012)

Fig. 14.6 (a) An open pit mining for Tertiary lignite in western Germany. The width of the pit is nearly 4 km with a depth of 120 m. The scene is at *50°52′N* and *6°20′E*. Partial refilling of the mine (the green terraces in the south) is now the action taken to minimize the consequences for the landscape. (b) Another lignite open pit mine in western Germany at *51°03′N* and *6°3′E*. The scene is 20 km wide. The partially refilled pit is 8 km wide and the hills of residual material is about 80 m high and about 1 km in diameter (Image credit: ©Google earth 2012)

Fig. 14.6 (continued)

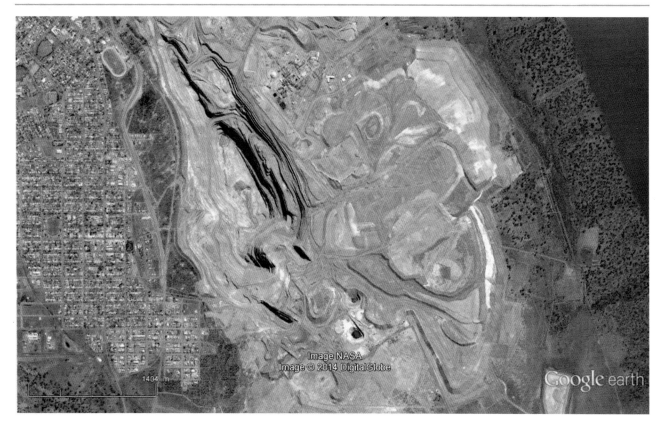

Fig. 14.7 Gold mine of Kalgoorlie, Western Australia, at *30°47′S* and *121°31′E*, in a 7 km wide scene. Mine is >200 m deep (max. 570 m). For mining of the gold, the removal of 85 million tons of rock without gold content is necessary per year (Image credit: ©Google earth 2012)

Fig. 14.8 Tom Price iron ore mine, Australia, at *22°46′S* and *117°46′E*. Scene is 10 km wide, showing greater than 40 km^2 of open mining up to 200 m deep; 28 million tons iron ore is moved per year (Image credit: ©Google earth 2012)

Fig. 14.9 Bingham Canyon open pit copper mine in Utah, USA, at *40°32′N* and *112°09′W*. Scene is about 15 km wide, showing the 8.5 km by 5.5 km open pit that is 600 m deep in the eastern end and 900 m deep in the western end. A huge landslide from the western wall with 165 million tons of rock occurred on April 11th, 2013 (Image credit: ©Google earth 2012)

Fig. 14.10 Copper mining in the Chuquicamata Mine (near Copiaco, Chile) at *27°31′S* and *70°18′W*. Scene is 7.5 km wide The maximum depth of the open pit is 1,000 m. It is the largest copper mine of the world, with >2.5 billion tons of copper ore mined during the last 100 years (Image credit: ©Google earth 2012)

Fig. 14.11 Deep shaft for diamond mining in old volcanic rocks in South Africa in Kimberley. The scene is centered at *28°44′S* and *24°46′E* and is 6 km wide. The pit is nearly 800 m of diameter (Image credit: ©Google earth 2012)

Fig. 14.12 Modification of slopes by road tracks in high mountain regions (St. Gotthard, old and new pass roads at the Switzerland-Italy border at *46°32′N* and *8°34′E*) (Image credit: ©Google earth 2012)

Fig. 14.13 Artificial shortening of a river course for shipping purposes (River Rhine in Germany at *49°13′N* and *8°24′E*) (Image credit: ©Google earth 2012)

Fig. 14.14 Modification of river runoff and sediment transport by reservoirs (dams indicated by arrows) around *47°05′N* and *119°37′W* along the Columbia River and the surrounding areas, USA (Image credit: ©Google earth 2012)

land reclamation along coastlines etc.), will result in a much faster transformation of the global surface than we have ever seen before.

With only a few images we will show some of the examples of anthropogenic landforms, which can be detected by a satellite and that are leaving a visible imprint on our natural environment. In many cases the consequences of these large-scale modifications have not been analyzed or modelled sufficiently. The discipline of "geomorphology" may help to understand and possibly mitigate associated risks better for the future of our natural environment.

Further Reading

Szabó J, Dávid L, Lóczy D (eds) (2010) Anthropogenic geomorphology – a guide to man-made landforms. Springer, Dordrecht

Index

A
A'a lava, 24, 28, 31, 34
Ablation, 296
Abrasion, 20, 245
Active layer, 175, 347, 349
Aeolian, 3, 255–290
Alluvial fan, 5, 76, 98, 99, 175, 176, 224, 236–238
Anabranching river, 199, 221–223
Andesite, 23
Antecedence, 88, 211
Anthropogenic, 164, 377–387
Anticline, 80, 83, 88, 90, 92, 93, 96, 108, 211
Ash, 18, 34, 36–38, 40, 47, 294, 296, 297
Asteroid, 123–125, 134
Asthenosphere, 4, 7, 18, 75
Atmosphere, 4, 113, 124, 125, 137, 184, 293
Auto-compaction, 236
Avalanche, 165, 168, 171, 173–175, 265, 317

B
Badland, 184, 187
Barchanoid, 268–270, 273, 284
Barchans, 261, 262, 268–270, 272, 273, 277
Basalt, 18, 22–24, 35–37, 49, 57, 71, 142, 187, 260
Batholith, 57, 143
Bed load, 184, 192, 194, 224, 377
Bifurcation, 236, 237, 342
Biosphere, 4, 294
Bomb, 36, 38, 131
Braided river, 195, 199, 217, 219, 224
Bulge, 113, 134, 135

C
Caldera, 18, 24, 28, 39, 41, 46, 47, 49, 69
Calving, 296, 299, 301, 304
Canyon, 138, 164, 194, 199, 200, 207, 230, 384
Cave, 24, 143, 147, 148, 151, 154, 161
Chemical weathering, 5, 126, 137–146, 192, 245
Cinder cone, 37–39, 45
Cirque, 41, 296, 307, 317, 326, 335, 349
Cockpit karst, 154, 158
Comet, 124, 125
Continental drift, 18
Convection, 4
Coral, 80, 154, 256
Corestone, 65, 112, 139, 142–145
Corrasion, 256, 261–263
Crater, 4, 20, 34, 37, 39–49, 51, 53, 90, 113, 123–135, 169
Crevasse, 114, 301, 306, 307
Cuesta, 28, 81, 92–96, 207, 209, 262

D
Dead ice, 317
Debris cone, 175, 347, 349
Debris fan, 217, 224, 236, 245
Debris flow, 167, 168, 173, 175, 176, 179
Deflation, 255–290
Delta, 113, 236, 242, 357, 358, 373, 377
Denudation, 88, 194, 353
Desertification, 256
Desert varnish, 138, 259
Desquamation, 139
Detersion, 317
Diatreme, 48–51
Dike, 52, 57–73, 187, 224, 377
Doline, 151
Drainage pattern, 3, 4, 42, 98
Drift, 4, 7, 18, 19, 35, 242, 256, 257, 273, 317
Drumlin, 334, 343
Dunes, 4, 154, 255–290
Dust storm, 256, 259

E
Earth pyramid, 337
Earthquake, 4, 5, 10–12, 18, 20, 81, 140, 164, 173–175, 317
Endogenic, 4, 137
Erg, 255, 286
Erosion, 3, 4, 19, 20, 42, 47, 48, 52, 65, 72, 80, 81, 90, 92, 98, 101, 103, 113, 114, 118, 124–126, 132, 135, 138, 139, 142, 161, 164, 167, 173, 184, 192, 194, 199, 203, 209, 224, 245, 255, 256, 263, 326, 331, 334, 370, 377
Erosive tools, 192
Escarpment, 164
Esker, 342, 344
Evaporation, 146, 161, 273, 347
Exfoliation, 65, 139, 142, 143, 166
Exogenic, 4, 5, 137
Extraterrestrial, 125, 128
Extrusion, 37, 44

F
Fanglomerate, 245
Fault, 4, 6, 7, 10, 75–77, 81–101, 115, 164, 171, 174, 199, 248, 337
Flash flood, 194, 245, 287
Fluvial, 3, 5, 99, 148, 168, 183–244, 331, 334, 344
Fold/folding, 4, 24, 75, 77, 80–92, 101, 114, 115, 207
Foraminifer, 80
Friction, 17, 124, 164, 166, 168, 175, 184, 194, 199, 256, 257, 260, 261, 306, 307, 311, 317, 350, 368
Frost shattering, 184, 224, 317, 331

Frost wedging, 138, 142, 164
Fumarole, 53, 54

G
Gelifluction, 224, 347, 353
Geologic window, 93
Geomorphology, 3–13, 199, 289, 294, 334, 347, 377, 387
Geosphere, 4
Geothermal, 17, 54, 55
Geyser, 54, 55
Glacial drift, 317
Glacial quarrying, 328, 331
Glacier
 budget, 317
 milk, 192
 mill, 306
 table, 317, 319
Glaciofluvial, 342
Glacis, 5, 245–253
Gooseneck, 213
Gorge, 5, 168, 194, 199, 201, 202, 213, 331
Gosses Bluff, 131
Graben, 6, 9, 11, 13, 19, 75–77, 97, 102, 238
Granite, 57, 65–67, 137–139, 141–145, 163, 166
Gravitaty, 5, 114, 163–181, 265 347, 349, 357
Groundwater, 48, 114, 141, 155, 161, 184, 269, 273, 277, 377

H
Hamada, 255
Honeycomb, 138, 143, 146
Hot spot, 6, 17, 19, 21, 23, 35, 37
Hydration, 142–143
Hydrolysis, 142–143
Hydrosphere, 4
Hydrostatic pressure, 202, 317
Hydrothermal, 54, 101

I
Ice age, 138, 202, 293, 352
Iceberg, 296, 301, 304
Ice wedge, 347, 349, 363, 364
Igneous, 23, 52, 57–73, 137, 141
Impact, 4, 5, 36, 90, 113, 114, 123–135, 141, 256, 294, 377
Inland ice, 306, 317, 334
Inselberg, 5, 245–253
Insolation, 142, 306, 317
Intrusion, 57–62, 67, 96, 113

J
Joint, 24, 52, 58, 60, 101–108, 111, 112, 138, 139, 142–145, 147–149, 151, 154, 161, 164, 207, 262, 331

K
Kames, 342, 344
Kar, 296, 317
Karren, 148, 149
Karst, 114, 147–161
Knickpoint, 82
Kopje, 246

L
Laminar, 184, 310, 317
Landslide, 5, 179–181, 384
Lapilli, 47
Lateral moraine, 310, 334, 340, 344
Lava, 6, 18–24, 28, 31, 34–39, 41, 44, 47, 54, 101, 142, 169, 175
Lava dome, 35, 37, 44
Lava tube, 24, 34
Lee-dune, 269, 277
Lithosphere, 4, 6, 7, 18–20, 75, 113, 125, 142, 236
Longitudinal dune, 269, 278

M
Maar, 48, 49, 51
Magma, 17–20, 23, 35, 37, 39, 48, 54, 57, 58, 112, 142
Mass movement, 5, 163–181
Meander, 196, 199, 203–207, 209, 211, 213, 215, 226
Medial moraine, 306, 310, 317, 320
Mega ripples, 284
Metamorphic, 61, 91, 138, 171
Meteor, 123, 124, 126
Meteorite, 114, 123–126, 131
Mid-Atlantic Ridge, 7, 19, 37
Mid-ocean ridge, 4, 6, 19, 23, 75
Mining, 377, 379, 381, 383–385
Mixed corrosion, 148
Mogote, 160
Moraine, 138, 217, 306, 308, 310, 317, 318, 320, 322, 324, 334, 337, 340, 342, 344
Morphometry, 3
Mudflow, 175
Muschelbruch, 331, 332
Mushroom rock, 256, 261

N
Natural levée, 224–226, 242
Near Earth Object (NEO), 123
Neck, 48, 52, 167, 199, 209, 213, 215
Nunatakker, 331, 334

O
Obsidian, 35
Ogive, 306, 311
Oser, 342, 344
Outlet glacier, 304, 317
Outwash plain, 246
Oxbow lake, 199, 204–207, 209
Oxidation, 142–143

P
Pahoehoe lava, 23, 24
Palsa, 347, 356
Parabolic dune, 261, 268, 273
Patterned ground, 347, 349, 350, 360, 361, 363
Pediment, 5, 245–253
Peneplain, 5, 245–253
Periglacial, 3, 347–373
Permafrost, 6, 173, 175, 347–373
Petrified dunes, 137
Physical weathering, 141–143, 199, 236, 256

Piedmont glacier, 296, 308
Pingo, 347, 357, 358
Planation surface, 246
Plateau glacier, 305, 306
Plate tectonics, 4, 18–20, 75, 124
Pleistocene, 51, 125, 293, 294, 306, 333, 334, 348
Pluton, 57, 59, 62, 64, 144
Polje, 151, 152, 154, 156, 157
Pollen, 294
Polygonal, 347, 349
Ponor, 154, 157
Post-volcanic, 49
Pothole, 184, 192, 317
Progradation, 236
Pumice, 38
Pyroclastic, 35–38, 54, 175

Q
Quasi-laminar, 184, 317
Quaternary, 3, 18, 126, 224, 293

R
Re-advance, 340
Reef, 154, 256
Reg, 255
Rhyolite, 23, 35
Rift, 6, 9, 17, 19, 23, 37, 75–77, 113
Ring of Fire, 4, 18, 19
Ripple marks, 261, 265
Roche moutonnée, 328
Rock avalanche, 165, 168, 171, 173, 174, 175
Rock fall, 5, 164–169, 349, 370
Rock glacier, 347, 349, 368
Rock pool, 53, 184
Rock slide, 168–173, 175, 370

S
Saltation, 184, 194, 255–257, 265
Salt dome, 90, 107, 108, 113, 115
Salt glacier, 114
Salt weathering, 138, 141, 142, 146
Sander, 342
Sculptural forms, 4
Sea floor spreading, 4, 6, 7
Seif dunes, 269
Sérac, 306
Serir, 255
Shadow weathering, 142
Shatter cone, 131–133
Shelf ice, 306
Shield volcano, 19, 22, 23, 36–37, 39, 41
Sickle dune, 331
Sill, 57, 71–73, 307, 317, 326
Sinkhole, 151, 152, 154–156, 324
Slump, 125, 168
Snow, 42, 175, 224, 293, 294, 296, 297, 306, 317, 348, 349
Snowball Earth, 293
Solifluction, 173, 224, 347, 349, 350, 352, 353
Solution, 3, 101, 114, 143, 146–148, 151, 154, 161, 192, 193
Speleothem, 161
Stalactite, 161

Stalagmite, 161
Star dune, 269, 284, 286, 287
Stone pavement, 255, 256, 259
Strained, 4, 199, 224
Stratovolcano, 35–37, 39, 42, 47
Striae, 317
Striation, 269, 273, 277, 317, 328, 331
String bog, 349, 366
Structural forms, 4
Subduction, 4, 6, 17, 18, 23, 35, 37, 57
Subsidence, 76, 113, 236
Surge, 124, 308, 317
Suspension, 192–194, 199, 224, 256, 317
Syncline, 80, 83, 90, 92

T
Tafoni, 143–146
Talus, 167
Talus cone, 166–169, 217
Tectonic, 3–6, 11, 18–20, 75–120, 124, 137, 142, 154, 171, 183, 184, 199, 246
Tephra, 36, 38, 297
Terminal moraine, 217, 308, 317, 334, 340, 342
Terrace, 5, 99, 125, 224, 229, 233, 235, 342, 347, 377–379, 381
Thermo-abrasion, 370
Thermokarst, 370, 371, 373
Thermophiles, 54
Thufur, 354, 356
Till, 317, 324, 337
Toma, 168, 170, 171
Tower karst, 157, 160
Transform fault, 6, 7, 76
Transversal dune, 168, 269
Tsunami, 39
Tuff, 36, 48, 296

U
Uvala, 152, 154
U-valley, 335

V
Ventifact, 256, 260
Volcano, 4, 6, 7, 17–24, 35–55, 57, 179, 294, 297, 305

W
Waterfall, 168, 198, 224
Weathering, 4, 5, 35, 65, 72, 91, 101, 103, 104, 112, 113, 124, 126, 131, 137–146, 154, 161, 163, 164, 166, 167, 180, 184, 192, 194, 199, 200, 224, 236, 245, 256, 317, 349

Y
Yardang, 257, 263

Z
Zabriskie Point, 187
Zagros Mountains, 90, 114